Solid State Physics

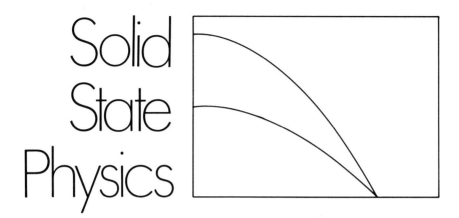

Solid State Physics

Harold T. Stokes

Brigham Young University

ALLYN AND BACON, INC.
Boston • London • Sydney • Toronto

Copyright © 1987 by Allyn and Bacon, Inc., 7 Wells Avenue, Newton, Massachusetts 02159. All rights reserved. No part of the material protected by this copyright notice may be reproduced or utilized in any form or by any means, electronic or mechanical, including photocopying, recording, or by any information storage and retrieval system, without written permission from the copyright owner.

Library of Congress Cataloging-in-Publication Data

Stokes, Harold T., 1947-
  Solid state physics.

  Bibliography: p.
  Includes index.
  1. Solid state physics. I. Title.
QC176.S784 1986   530.4'1   86-22213
ISBN 0-205-10508-4

Printed in the United States of America.

10 9 8 7 6 5 4 3 2 1   90 89 88 87 86

CONTENTS

Preface

Chapter 1. Crystal Structure

1-1	Introduction	1
1-2	Lattices	2
1-3	Basis Vectors	3
1-4	Simple Cubic Lattice	6
1-5	Unit Cells	6
1-6	Crystal Directions and Planes	8
1-7	Body-Centered Cubic Lattice	12
1-8	Face-Centered Cubic Lattice	17
1-9	Sodium Chloride Structure	21
1-10	Cesium Chloride Structure	25
1-11	Zincblende Structure	26
1-12	Diamond Structure	28
1-13	Other Bravais Lattices	30
1-14	Ionic Bond	32
1-15	Covalent Bond	34
1-16	Metallic Bond	35

Chapter 2. X-Ray Diffraction

2-1	Waves	37
2-2	Interference	39
2-3	Multiple-Slit Diffraction	42
2-4	X-Ray Diffraction in Crystals	44
2-5	Diffraction from Real Atoms	49
2-6	Reciprocal Lattice	51
2-7	Experimental Methods	57

Chapter 3. Lattice Vibrations

3-1	Harmonic Motion	61
3-2	One-Dimensional Monatomic Lattice	63
3-3	First Brillouin Zone	68
3-4	One-Dimensional Diatomic Lattice	72
3-5	Three-Dimensional Crystals	79
3-6	Thermal Expansion	84

Chapter 4. Classical Model of Metals

- 4-1 Conduction Electrons 89
- 4-2 Electric Current 90
- 4-3 Conductivity 92
- 4-4 Hall Effect 95
- 4-5 Cyclotron Resonance 97

Chapter 5. Waves and Particles

- 5-1 Photons 99
- 5-2 Phonons 101
- 5-3 Inelastic Scattering of Neutrons 102
- 5-4 Inelastic Scattering of Photons 106
- 5-5 Wave-like Properties of Particles 111

Chapter 6. Quantum Mechanics

- 6-1 Wave Functions 115
- 6-2 Schroedinger's Equation 117
- 6-3 Wave Function of a Free Particle 118
- 6-4 Particle in a Box 125
- 6-5 Tunneling 128
- 6-6 Wave Functions in Three Dimensions 130

Chapter 7. Free-Electron Quantum Model of Metals

- 7-1 Particle in a Box 133
- 7-2 Periodic Boundary Conditions 135
- 7-3 Density of States 136
- 7-4 Pauli's Exclusion Principle 138
- 7-5 Fermi-Dirac Distribution Function 142
- 7-6 Electrical Conductivity 147

Chapter 8. Band Theory of Metals

- 8-1 Interaction with Ions 153
- 8-2 Bloch Functions 157
- 8-3 Three-Dimensional Crystals 158
- 8-4 Number of Electron States in a Band 162
- 8-5 Fermi Surface 166
- 8-6 Atomic Model 166

Chapter 9. Electrical Conductivity of Metals

- 9-1 Group Velocity 171
- 9-2 Equation of Motion 173
- 9-3 Electrical Conductivity 174
- 9-4 Metals, Insulators, Semiconductors 178
- 9-5 Electron Collisions 180
- 9-6 Effective Mass 183
- 9-7 Holes 185
- 9-8 Hall Effect 189
- 9-9 Cyclotron Resonance 190

Chapter 10. Semiconductors

- 10-1 Band Structure 191
- 10-2 Simple Model 191
- 10-3 Density of Electrons and Holes 194
- 10-4 Physical Picture 200
- 10-5 Doped Semiconductors 201
- 10-6 Temperature Dependence of n 206
- 10-7 Electrical Conductivity 206
- 10-8 Hall Effect 210
- 10-9 Band Structure of Real Semiconductors . . . 211

Chapter 11. p-n Junctions in Semiconductors

- 11-1 The Junction 215
- 11-2 Diffusion of Electrons and Holes 215
- 11-3 Electric Field and Contact Potential 218
- 11-4 Depletion Layer 218
- 11-5 Fermi Level: Calculation of Contact Potential 221
- 11-6 Width of Depletion Layer 224
- 11-7 Currents across Junction in Equilibrium . . . 228
- 11-8 Biased Junctions 231
- 11-9 Capacitance of Junction 238

Chapter 12. Semiconductor Devices

- 12-1 Diode 241
- 12-2 Bipolar Junction Transistor 245
- 12-3 Field-Effect Transistor 249

12-4	Metal-Semiconductor Junctions	254
12-5	Optical Absorption	258
12-6	Photosensitive Devices	260
12-7	Light-Emitting Diode	262
12-8	Lasers	266

Chapter 13. Superconductivity

13-1	Introduction	273
13-2	Trapped Magnetic Flux and Persistent Currents	274
13-3	Meissner Effect	275
13-4	Penetration of Magnetic Fields	276
13-5	Critical Fields	278
13-6	Type II Superconductors	279
13-7	BCS Theory	281
13-8	Isotope Effect	288
13-9	Absorption of Electromagnetic Radiation	289
13-10	Tunneling	290

Appendix 1.	Some Physical Constants	293
Appendix 2.	The Elements	294
Appendix 3.	Crystal Structures	298
Appendix 4.	Properties of Metals	300
Appendix 5.	Properties of Semiconductors	302
Appendix 6.	Impurity Levels in Silicon and Germanium	303
Appendix 7.	Properties of Superconductors	304
Appendix 8.	Units	305
Appendix 9.	Further Reading	308
Index		310

Preface

I wrote this book for the course, Principles of Solid State Physics, developed at Brigham Young University. The course was specifically intended for non-physics majors. The prerequisites are two semesters of introductory physics that cover mechanics and electric and magnetic fields. A course in modern physics is *not* a prerequisite. Thus, many topics of modern physics, such as quantum mechanics, are included in this course to the extent that they are needed to understand the aspects of solid state physics covered.

The purpose of this course is to acquaint the student with the fundamental physics of solids. This emphasis is on understanding the behavior of electrons in metals and semiconductors. In the first 3 chapters, crystal structure, x-ray diffraction, and lattice vibrations are discussed as a natural way to introduce concepts such as translational symmetry, wave interference, reciprocal lattice, and the first Brillouin zone. Also, two chapters (5–6) on quantum theory are included to teach the basic ideas and principles of quantum mechanics needed. All of these concepts are brought together to form an understandable modern picture of electrons in metals (chapters 7–9), in semiconductors (chapters 10–12), and in superconductors (chapter 13). Numerous examples, problems, and figures are used to illustrate the material being presented. There are 181 problems and 188 figures in this book.

The reader may note that some topics usually found in conventional solid state physics textbooks are missing from this book, such as heat capacity, magnetism, dielectrics, etc. This is a one-semester course. I wanted to treat a few central topics in great detail rather than describe superficially a great many topics.

I wish to thank Professor Dean Barnett and Professor William Evenson for critically reading portions of this book and for the many valuable suggestions they have made.

Solid State Physics

CHAPTER 1

CRYSTAL STRUCTURE

1-1 Introduction

Many solids around us are crystalline. Obvious examples are diamonds and other precious stones which have an outward crystalline appearance. Many metals are also crystalline. However, they are usually composed of numerous small crystals fused together so that their outward appearance is *not* crystalline. In this book, we will consider only crystalline solids.

A *crystal* is a solid in which all the atoms are arranged in a periodic manner. As a simple example, consider the cubic arrangement of atoms shown in Fig. 1-1. We show only a portion of the crystal. We imagine that it extends out in all directions to infinity. Real crystals, of course, have finite dimensions, but,

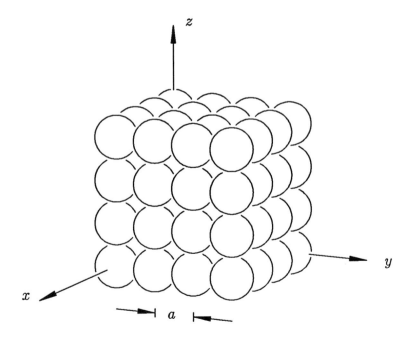

Fig. 1-1. Simple cubic arrangement of atoms.

for now, we consider the crystal to be infinitely large with no surfaces.

The atoms in the crystal shown in Fig. 1-1 are in **equivalent positions**. If we sit on one of the atoms, we cannot tell where we are by looking at the neighboring atoms. All atoms have exactly the same surroundings. (This, of course, is only strictly true in an *infinite* crystal.) If we move the entire crystal in some direction so that each atom is now at a position where some other atom used to be, the crystal looks the same as before. We cannot tell that it has been moved. This is called **translational symmetry**. We may now state the definition of a crystal more precisely. A **crystal** is a solid which has translational symmetry.

1-2 Lattices

In order to quantitatively describe a crystal, we introduce a group of geometric points called the crystal **lattice** which defines the positions of the atoms. As an example, consider the two-dimensional square lattice shown in Fig. 1-2. This lattice is a set of geometric points on a plane. If we were to place

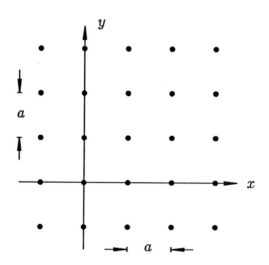

Fig. 1-2. The two-dimensional square lattice.

an atom at each point, then we would have a two-dimensional crystal. All lattice points in the figure are equivalent. This lattice has translational symmetry in two dimensions.

1-3 Basis Vectors

A **lattice vector** is a vector which takes us from one lattice point to any other lattice point. Obviously, all lattice vectors **R** in the two-dimensional square lattice (Fig. 1-2) have the form

$$\mathbf{R} = n_1 a \hat{\imath} + n_2 a \hat{\jmath}, \tag{1-1}$$

where n_1 and n_2 are integers (including negative values and zero), and a is the distance between adjacent lattice points in the x or y directions, as shown in Fig. 1-2. $\hat{\imath}$ and $\hat{\jmath}$ are unit vectors in the x and y directions, respectively. If we define two vectors (see Fig. 1-3),

$$\begin{aligned} \mathbf{a}_1 &= a\hat{\imath}, \\ \mathbf{a}_2 &= a\hat{\jmath}, \end{aligned} \tag{1-2}$$

then we can write **R** as

$$\mathbf{R} = n_1 \mathbf{a}_1 + n_2 \mathbf{a}_2. \tag{1-3}$$

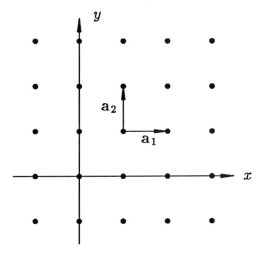

Fig. 1-3. Basis vectors for the square lattice.

Thus, *any* lattice vector **R** can be written as a linear combination of $\mathbf{a}_1$ and $\mathbf{a}_2$ (using integers, n_1 and n_2). Conversely, it is also true that any linear combination of $\mathbf{a}_1$ and $\mathbf{a}_2$ (using integers, n_1 and n_2) is a lattice vector **R**. Such vectors, $\mathbf{a}_1$ and $\mathbf{a}_2$, are called **basis vectors** of the lattice.

The choice of basis vectors, $\mathbf{a}_1$ and $\mathbf{a}_2$, is not unique. We could just as well choose (see Fig. 1-4)

$$\mathbf{a}'_1 = a\hat{\imath}, \qquad (1\text{-}4)$$
$$\mathbf{a}'_2 = a\hat{\imath} + a\hat{\jmath}.$$

For example, consider the lattice vector $\mathbf{R} = a\hat{\imath} + 2a\hat{\jmath}$. This can be written as $\mathbf{R} = \mathbf{a}_1 + 2\mathbf{a}_2$ or as $\mathbf{R} = -\mathbf{a}'_1 + 2\mathbf{a}'_2$ as shown in Fig. 1-5.

There are an infinite number of ways to choose basis vectors for a given lattice. There is, however, usually a **conventional** choice of basis vectors. For example, the conventional basis vectors for the square lattice are those given in Eq. (1-2). Basis vectors can be found for any lattice of *equivalent* points.

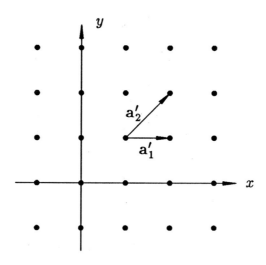

Fig. 1-4. An alternate choice of basis vectors for the square lattice.

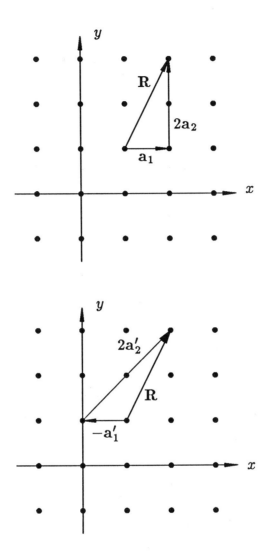

Fig. 1-5. Lattice vector **R** expressed as a linear combination of the basis vectors, a_1 and a_2, and also as a linear combination of the basis vectors, a_1' and a_2'.

1-4 Simple Cubic Lattice

The extension to three dimensions is straightforward. The lattice which underlies the crystal structure in Fig. 1-1 has basis vectors given by

$$\begin{aligned} \mathbf{a}_1 &= a\hat{\imath}, \\ \mathbf{a}_2 &= a\hat{\jmath}, \\ \mathbf{a}_3 &= a\hat{k}, \end{aligned} \quad (1\text{-}5)$$

and the lattice vector is given by

$$\mathbf{R} = n_1\mathbf{a}_1 + n_2\mathbf{a}_2 + n_3\mathbf{a}_3. \quad (1\text{-}6)$$

This lattice is called **simple cubic** (sc).

1-5 Unit Cells

A crystal can always be divided into "building blocks" called **unit cells**. Each unit cell has the same shape, the same volume, and the same contents. For the sc lattice, we may choose the unit cell to be a cube of side a (see Fig. 1-6). The

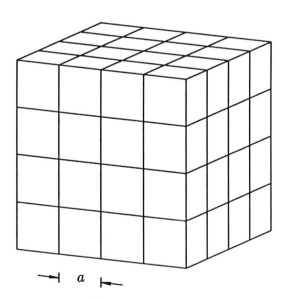

Fig. 1-6. Unit cells of the sc lattice

CHAPTER 1 CRYSTAL STRUCTURE

choice of position of the lattice point within the unit cell is arbitrary. We may arrange the cubes so that each cube contains one lattice point at its center (see Fig. 1-7a). Alternately, we may arrange the cubes so that the lattice points are at the corners of the cube (see Fig. 1-7b). The second choice is the **conventional unit cell**.

Each of these unit cells contains one lattice point. This is obviously true for the unit cell in Fig. 1-7a which contains one lattice point in the center. But the conventional unit cell in Fig. 1-7b appears to contain *eight* lattice points, one at each corner. Actually, each of those lattice points is shared by eight neighboring unit cells that adjoin at the corner so that each unit cell contains "$\frac{1}{8}$ lattice point" at the corner. Eight such lattice points give us a total of one lattice point in the unit cell.

Just as the choice of basis vectors for a lattice is not unique, the choice of unit cells is also not unique. For example, we could just as well choose the unit cell shown in Fig. 1-8. We only require that the unit cells be identical to each other and fill all space. There are an infinite number of ways to choose the unit cell. However, the *conventional* unit cell for the sc

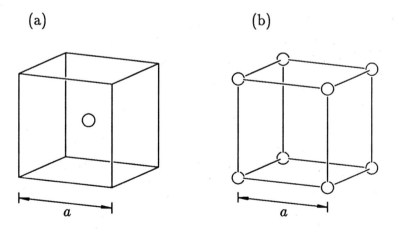

Fig. 1-7. The unit cell for the sc lattice (a) with a lattice point at the center and (b) with a lattice point at each corner (conventional unit cell).

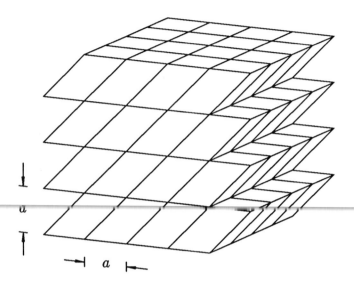

Fig. 1-8. An alternate choice of unit cell for the sc lattice.

lattice is the cube shown in Fig. 1-7b. The distance a between adjacent atoms in the x, y, or z direction is called the **lattice parameter**.

1-6 Crystal Directions and Planes

Directions in crystals are usually represented in shorthand by three integers inside a set of square brackets. The direction $\mathbf{R} = n_1 a \hat{\imath} + n_2 a \hat{\jmath} + n_3 a \hat{k}$ in a cubic crystal, for example, is written as $[n_1, n_2, n_3]$. The integers are usually chosen to be as small as possible. Three common directions in cubic crystals with which we will deal are (see Fig. 1-9)

$$\begin{aligned}
\mathbf{R} &= a\hat{\imath} & \text{or} \quad & [100], \\
\mathbf{R} &= a\hat{\imath} + a\hat{\jmath} & \text{or} \quad & [110], \\
\mathbf{R} &= a\hat{\imath} + a\hat{\jmath} + a\hat{k} & \text{or} \quad & [111],
\end{aligned} \qquad (1\text{-}7)$$

Consider a hypothetical crystal which has one atom at each lattice point of an sc lattice with $a = 5.00$ Å. (No naturally occurring element forms an sc lattice.) Starting from an atom

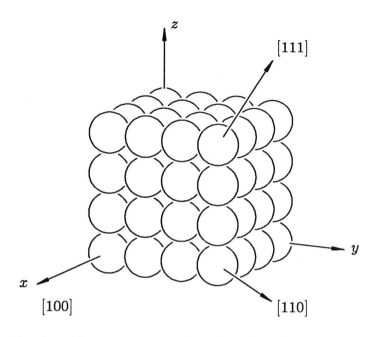

Fig. 1-9. Three common directions in cubic crystals.

at the origin, we see that along the [100] direction, there are atoms at $a\hat{\imath}$, $2a\hat{\imath}$, $3a\hat{\imath}$, etc. The distance between adjacent atoms along the [100] direction is $a = 5.00$ Å, which is the length of the vector $\mathbf{a}_1 = a\hat{\imath}$.

If we go along the [110] direction from the atom at the origin, we find atoms at $a\hat{\imath} + a\hat{\jmath}$, $2a\hat{\imath} + 2a\hat{\jmath}$, $3a\hat{\imath} + 3a\hat{\jmath}$, etc., and the distance between adjacent atoms is the length of the vector $a\hat{\imath} + a\hat{\jmath}$, which is $\sqrt{2}a = 7.07$ Å. Similarly, along the [111] direction, atoms are at $a\hat{\imath} + a\hat{\jmath} + a\hat{k}$, $2a\hat{\imath} + 2a\hat{\jmath} + 2a\hat{k}$, $3a\hat{\imath} + 3a\hat{\jmath} + 3a\hat{k}$, etc., and the distance between adjacent atoms is the length of the vector $a\hat{\imath} + a\hat{\jmath} + a\hat{k}$, which is $\sqrt{3}a = 8.66$ Å.

We would find, in general, that along any given direction in a crystal, atoms are evenly spaced. The distance between adjacent atoms is smallest along the [100] direction. These atoms are **nearest neighbors** to each other. If we imagine the atoms to be "hard" spheres such as those shown in Fig. 1-1, then we see that the nearest neighbors "touch" each other, and

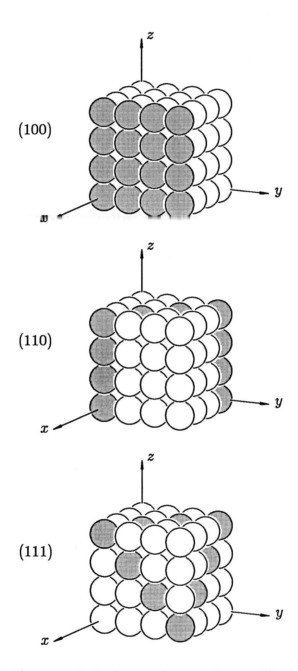

Fig 1-10. Some typical planes of atoms in a cubic crystal. The atoms forming the planes are shaded.

the distance between the centers of nearest-neighbor atoms is the diameter of the atoms. Thus, in this hypothetical crystal, the diameter of the atoms is 5.00 Å.

Planes of atoms in a crystal are usually represented in shorthand by three integers inside a set of *parentheses*. In cubic crystals, a plane denoted by (n_1, n_2, n_3) is perpendicular to the direction $[n_1, n_2, n_3]$. These three integers n_1, n_2, n_3, when referring to crystal planes, are called **Miller indices**. (In non-cubic crystals, the meaning of Miller indices is more complicated.) In Fig. 1-10 are shown some typical planes in a

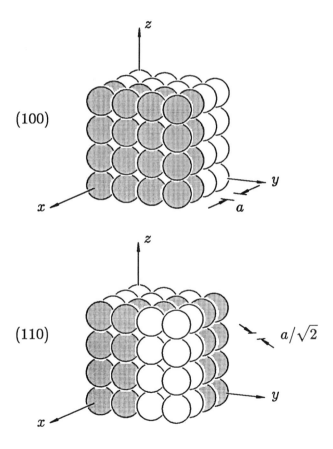

Fig 1-11. Distances between adjacent planes of atoms for two cases in a cubic crystal. The adjacent planes of atoms are shaded.

cubic crystal. Distances between adjacent planes of atoms can be computed by inspection for the simpler cases. For example, from Fig. 1-11, we see that the distance between adjacent (100) planes is a and between adjacent (110) planes is $a/\sqrt{2}$. For our hypothetical crystal of $a = 5.00$ Å, these distances are 5.00 Å and 3.54 Å, respectively.

The atomic density n of a crystal (in units of atoms/unit volume) is easily obtained by considering a single unit cell. For our hypothetical crystal, there is one atom per unit cell, and the volume of the unit cell is a^3. Thus, $n = a^{-3}$. For $a = 5.00$ Å, we have $n = 8.00 \times 10^{21}$ atoms/cm^3. To obtain the *mass* density ρ (in units of g/cm^3, for example), we only need to know the mass of each atom. We will illustrate this later for the case of an actual crystal.

1-7 Body-Centered Cubic Lattice

We can form a new lattice which is different from the sc lattice by placing an additional lattice point at the center of the unit cell of Fig. 1-7b. The resulting lattice is called body-centered cubic (bcc) and is shown in Fig. 1-12. Note that every lattice point is equivalent to every other lattice point. Each of the original sc lattice points is also in the body-centered position of eight of the new lattice points.

The conventional unit cell for the bcc lattice is a cube of side a as shown in Fig. 1-13. We see that there are two lattice points in this unit cell: a lattice point in the center and $\frac{1}{8}$ lattice point at each of the eight corners. This is *not* the smallest possible unit cell which can be constructed for the bcc lattice. The smallest possible unit cell, called the **primitive unit cell**, contains only *one* lattice point. Its volume is *half* the volume of the conventional unit cell shown in Fig. 1-13.

The choice of primitive unit cell is not unique. There are an infinite number of possible ways to choose it. However, in physics, we usually use the primitive unit cell called the **Wigner-Seitz cell**. This cell contains one lattice point at its center and contains the region of space that is closer to that point than to any other lattice point. This is best explained by example. The Wigner-Seitz cell for a two-dimensional square

CHAPTER 1 CRYSTAL STRUCTURE   13

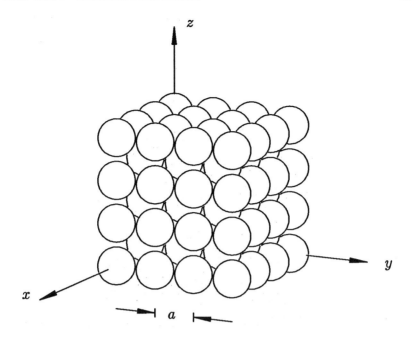

Fig. 1-12. Body-centered cubic arrangement of atoms.

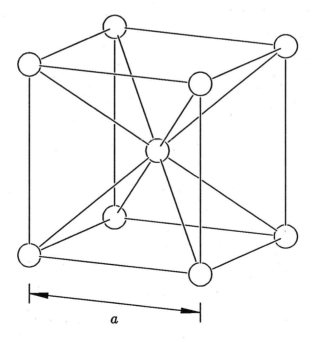

Fig. 1-13. The conventional unit cell of the bcc lattice

lattice is shown in Fig. 1-14. For the two-dimensional centered-rectangular lattice shown in Fig. 1-15, the result is more complicated. Note that each line segment which forms a cell boundary is a perpendicular bisector of a line joining two lattice points.

In three dimensions, the case of the sc lattice is simple. The Wigner-Seitz cell is the cube we already showed in Fig. 1-7a. For the bcc lattice, the Wigner-Seitz cell is a "truncated octahedron" as shown in Fig. 1-16. Its volume is $\frac{1}{2}a^3$. These cells nest together and fill all space.

The basis vectors of a bcc lattice are *not* those given in Eq. (1-5) for the sc lattice. No linear combination of those vectors (using integers) can take us to any of the body-centered lattice points such as $\mathbf{R} = \frac{1}{2}a\hat{i} + \frac{1}{2}a\hat{j} + \frac{1}{2}a\hat{k}$. There is no unique way to choose the basis vectors of a bcc lattice, but a very common choice is

$$\begin{aligned}
\mathbf{a}_1 &= -\tfrac{1}{2}a\hat{i} + \tfrac{1}{2}a\hat{j} + \tfrac{1}{2}a\hat{k}, \\
\mathbf{a}_2 &= \tfrac{1}{2}a\hat{i} - \tfrac{1}{2}a\hat{j} + \tfrac{1}{2}a\hat{k}, \\
\mathbf{a}_3 &= \tfrac{1}{2}a\hat{i} + \tfrac{1}{2}a\hat{j} - \tfrac{1}{2}a\hat{k}.
\end{aligned} \quad (1\text{-}8)$$

All lattice points can be expressed in the form $\mathbf{R} = n_1\mathbf{a}_1 + n_2\mathbf{a}_2 + n_3\mathbf{a}_3$. For example, $\frac{1}{2}a\hat{i} + \frac{1}{2}a\hat{j} + \frac{1}{2}a\hat{k} = \mathbf{a}_1 + \mathbf{a}_2 + \mathbf{a}_3$. Also, as another example, $a\hat{i} = \mathbf{a}_2 + \mathbf{a}_3$.

Directions and planes in a bcc crystal are *not* labeled in reference to these new basis vectors, but are labeled according to the convention we introduced for the sc lattice. Thus, directions $[n_1, n_2, n_3]$ and planes (n_1, n_2, n_3) have the same meaning with respect to the x, y, and z axes as in the sc lattice shown in Figs. 1-9 and 1-10. In the bcc lattice, points along the [111] direction are nearest neighbors. The distance between them is the length of the vector $\mathbf{R} = \frac{1}{2}a\hat{i} + \frac{1}{2}a\hat{j} + \frac{1}{2}a\hat{k}$ which is $\frac{1}{2}a\sqrt{3} = 0.866a$.

Examples of elements which form bcc crystals are given in Appendix 3. For example, iron (Fe) forms a bcc crystal with $a = 2.86$ Å. The distance between nearest-neighbor atoms is $0.866a = 2.48$ Å. One can think of this as the "diameter" of the Fe atom. The easiest way to compute the atom density n

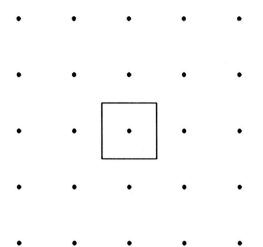

Fig. 1-14. The Wigner-Seitz cell for a two-dimensional square lattice.

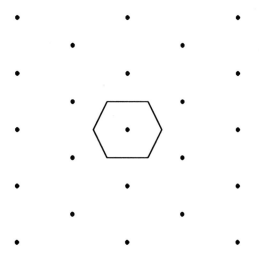

Fig. 1-15. Wigner-Seitz cell for the two-dimensional centered-rectangular lattice.

(a)

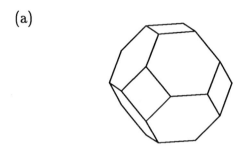

(b)

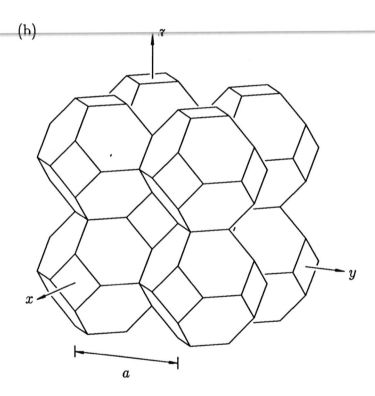

Fig. 1-16. (a) The Wigner-Seitz cell for the bcc lattice. (b) These cells fit together to fill all space.

of Fe is to use the conventional unit cell of Fig. 1-13. It contains two atoms and has a volume of a^3. Thus, $n = 2a^{-3} = 8.55 \times 10^{22}$ atoms/cm^3. To obtain the mass density ρ, we need to know the mass of a single Fe atom. The **atomic mass** given in Appendix 2 is the mass in units of atomic mass units (amu). From Appendix 2, we find the mass of one Fe atom is 55.847 amu or 9.27×10^{-23} g. Thus, the mass density of Fe is $\rho = (8.55 \times 10^{22}$ atoms/cm$^3) \times (9.27 \times 10^{-23}$ g/atom$) = 7.93$ g/cm^3.

Problem 1-1. Consider a crystal of iron (Fe). Find the distance between adjacent atoms in the [100] direction. Repeat for the [110] and [111] directions. Find the distance between the (100) planes. Repeat for the (110) planes. Answer: 2.86 Å, 4.04 Å, 2.48 Å, 1.43 Å, 2.02 Å.

Problem 1-2. Using the data in Appendix 3, find the atomic diameter of lithium (Li). Repeat for sodium (Na), potassium (K), rubidium (Rb), and cesium (Cs). (Consider the atoms to be spheres which touch each other.) Answer: 3.03 Å, 3.72 Å, 4.50 Å, 4.84 Å, 5.63 Å.

Problem 1-3. Using data in Appendix 3, find the volume of a primitive unit cell in a crystal of chromium (Cr). Answer: 11.8 Å^3.

1-8 Face-Centered Cubic Lattice

Yet another lattice can be formed from the sc lattice by placing a lattice point at the center of each *face* of the unit cell shown in Fig. 1-7b. The resulting lattice is called face-centered cubic (fcc) and is shown in Fig. 1-17. As with the bcc lattice, every lattice point in the fcc lattice is equivalent. The *conventional* unit cell for the fcc lattice is a cube of side a as shown in Fig. 1-18. Let us count the number of lattice points in this unit cell. There are eight corners (each containing $\frac{1}{8}$ lattice point) and six faces (each containing $\frac{1}{2}$ lattice point) which gives us a total of four lattice points. This, of course, is

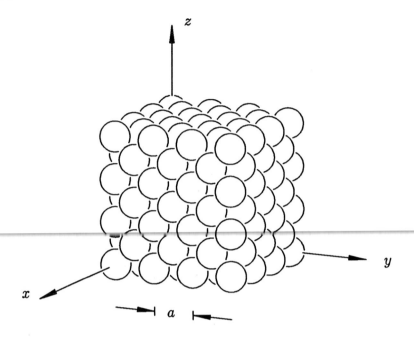

Fig. 1-17. Face-centered cubic arrangement of atoms.

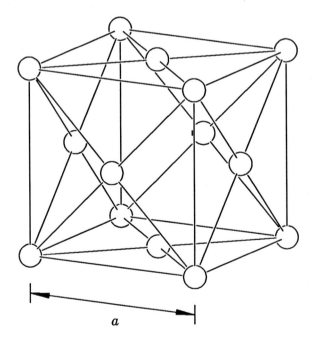

Fig. 1-18. The conventional unit cell of the fcc lattice

CHAPTER 1 CRYSTAL STRUCTURE   19

not a primitive unit cell for the fcc lattice. The Wigner-Seitz primitive unit cell for the fcc lattice is a rhombic dodecahedron and is shown in Fig. 1-19. It contains one lattice point, and thus its volume is $\frac{1}{4}a^3$.

A common choice for basis vectors of the fcc lattice is the following:

$$\begin{aligned} \mathbf{a}_1 &= \tfrac{1}{2}a\hat{j} + \tfrac{1}{2}a\hat{k} \\ \mathbf{a}_2 &= \tfrac{1}{2}a\hat{i} + \tfrac{1}{2}a\hat{k} \\ \mathbf{a}_3 &= \tfrac{1}{2}a\hat{i} + \tfrac{1}{2}a\hat{j} \end{aligned} \qquad (1\text{-}9)$$

Lattice points along the [110] direction are nearest neighbors. The distance between them is the length of the vector $\tfrac{1}{2}a\hat{i}+\tfrac{1}{2}a\hat{j}$ which is $\tfrac{1}{2}a\sqrt{2} = 0.707a$. Examples of elements which form fcc crystals are given in Appendix 3.

Problem 1-4. Consider a crystal of copper (Cu). Find the distance between adjacent atoms in the [100] direction. Repeat for the [110] and [111] directions. Find the distance between the (100) planes. Repeat for the (110) planes. Answer: 3.61 Å, 2.55 Å, 6.25 Å, 1.81 Å, 1.28 Å.

Problem 1-5. Using the data in Appendix 3, find the atomic diameter of copper (Cu). Repeat for silver (Ag) and gold (Au). (Consider the atoms to be spheres which touch each other.) Answer: 2.55 Å, 2.88 Å, 2.88 Å.

Problem 1-6. Using data in Appendices 2 and 3, find the density (in g/cm^3) of aluminum (Al). Answer: 2.72 g/cm^3.

Problem 1-7. The structure of iron is found to be bcc below 910°C and fcc above 910°C. The density of iron increases by 1.0% as it goes through the transition from bcc to fcc at 910°C. By what percentage does the nearest-neighbor distance between iron atoms change? Answer: 2.5%.

Problem 1-8. Using data in Appendix 3, find the volume of a primitive unit cell in a crystal of nickel (Ni). Answer: 10.9 Å^3.

(a)

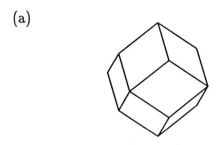

(b)

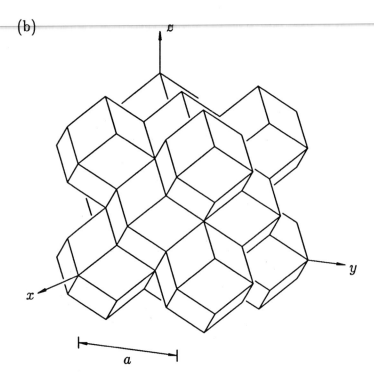

Fig. 1-19. (a) The Wigner-Seitz cell for the fcc lattice. (b) These cells fit together to fill all space.

Problem 1-9. $\mathbf{R} = a\hat{\imath}$ is a lattice vector of the fcc lattice. Write this vector in the form $\mathbf{R} = n_1\mathbf{a}_1 + n_2\mathbf{a}_2 + n_3\mathbf{a}_3$, using the basis vectors in Eq. (1-9).

Problem 1-10. Consider a box of volume 1.00 m^3 filled with balls of diameter 1.00 mm. (a) If we pack the balls in an sc lattice, how many can we get into the box? (b) Repeat for bcc lattice. (c) Repeat for fcc lattice. Answer: 1.00×10^9, 1.30×10^9, 1.41×10^9.

1-9 Sodium Chloride Structure

Next, let us consider the structure of NaCl (sodium chloride, common table salt). In Fig. 1-20, we see that the Na and Cl atoms occupy the lattice points of an sc lattice. However, since some lattice points are occupied by Na atoms and others by Cl atoms, not all lattice points are *equivalent*. We can tell

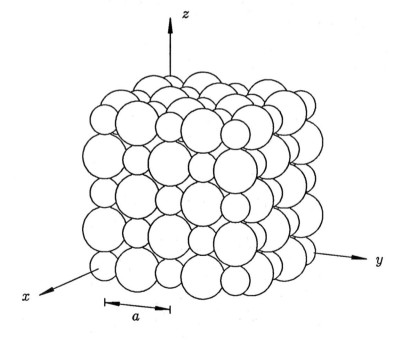

Fig. 1-20. The sodium chloride structure. The small spheres represent Na atoms, and the large spheres represent Cl atoms.

the difference between the two kinds of sites because the atoms there are different.

If we look carefully at Fig. 1-20, we can see that the Na sites form a set of lattice points which *are* equivalent. We cannot tell the difference between Na atoms. They are each in identical surroundings. Such a set of lattice points which are equivalent to each other is called a **Bravais lattice**. The Bravais lattice of NaCl corresponds to the set of Na sites and is fcc as we can see in Fig. 1-20. Actually, the absolute location of the lattice points is arbitrary, and we could just as well put the lattice points at the Cl sites, which are also equivalent to each other, or we could put the lattice points between Na and Cl atoms. The choice of origin for the lattice is arbitrary. For any given origin $\mathbf{R}_0$ for the lattice, all points $\mathbf{R}_0 + n_1 \mathbf{a}_1 + n_2 \mathbf{a}_2 + n_3 \mathbf{a}_3$ (where $\mathbf{a}_1$, $\mathbf{a}_2$, and $\mathbf{a}_3$ are basis vec-

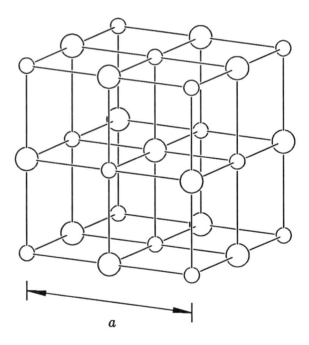

Fig. 1-21. The conventional unit cell for the sodium chloride structure. The large and small spheres represent two different types of atoms.

tors for the fcc Bravais lattice) are at equivalent positions in the crystal.

Associated with each lattice point are two atoms, Na and Cl. These two atoms are called the **basis**. (Do not confuse this usage of the word *basis* with that in *basis vector*.) The lattice is a set of geometric points. The basis is a set of one or more real atoms associated with each lattice point. A *crystal* is a basis combined with a lattice.

The conventional unit cell for the NaCl structure is the same as that for its fcc Bravais lattice and is shown in Fig. 1-21. This unit cell contains four Na atoms and four Cl atoms. We can obtain this result by actually counting them as we did for the bcc and fcc lattices, but the result is more easily obtained by remembering that the conventional unit cell of the fcc lattice contains four lattice points. Since the basis for the NaCl structure is one Na atom and one Cl atom, then the conventional unit cell must contain four of these bases, or four Na and four Cl atoms. The Wigner-Seitz primitive unit cell for the NaCl structure is that of its Bravais fcc lattice, the rhombic dodecahedron in Fig. 1-19, and, of course, contains one Na atom and one Cl atom. Examples of crystals with the NaCl structure are given in Appendix 3.

Problem 1-11. Find the distance between nearest-neighbor atoms in sodium chloride (NaCl). Answer: 2.82 Å.

Problem 1-12. The density of sodium chloride (NaCl) is 2.165 g/cm^3. Using the atomic masses in Appendix 2, calculate the lattice parameter a and compare with the value given in Appendix 3.

Problem 1-13. Using data in Appendix 2, find the volume of a primitive unit cell in a crystal of potassium chloride (KCl). Answer: 61.6 Å^3.

Problem 1-14. Consider the two-dimensional crystal shown in Fig. 1-22. The o and • symbols represent two different kinds of atoms. Are all o atoms in equivalent positions? Are all

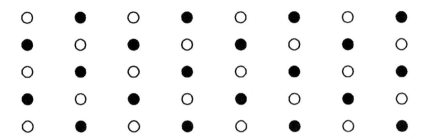

Fig. 1-22. Two-dimensional crystal for Problem 1-14.

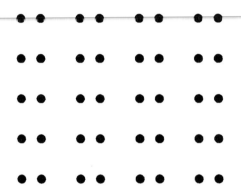

Fig. 1-23. Two-dimensional crystal for Problem 1-15.

• atoms in equivalent positions? Draw a "conventional" unit cell which is rectangular in shape and has a ○ atom on each corner. How many ○ atoms are in this unit cell? How many • atoms? Draw a Wigner-Seitz primitive unit cell centered on a ○ atom.

Problem 1-15. Consider the two-dimensional crystal in Fig. 1-23. There is only one kind of atom in the crystal, represented by the symbol • in the figure. Are all atoms in equivalent positions? Draw a primitive unit cell. How many atoms are in this cell?

1-10 Cesium Chloride Structure

Another rather common structure is that of CsCl (cesium chloride, see Fig. 1-24). The Cs and Cl atoms sit at the lattice points of a bcc lattice, but, as with NaCl, since the sites are occupied by different atoms and are not equivalent, the Bravais lattice is given by the set of lattice points just occupied by the Cs atoms and is therefore sc. The basis associated with each lattice point is one Cs atom and one Cl atom. The conventional unit cell of CsCl is the same as that of its sc Bravais lattice (see Fig. 1-25) and contains one Cs atom and one Cl atom. Examples of crystals with the CsCl structure are given in Appendix 3.

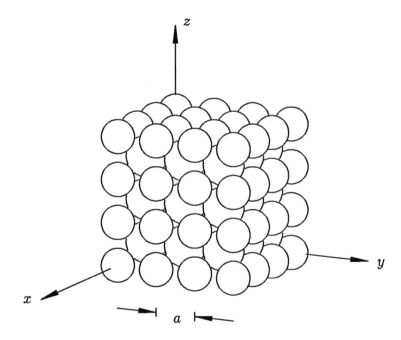

Fig. 1-24. The cesium chloride structure. The small spheres represent the Cs atoms and the large spheres represent the Cl atoms.

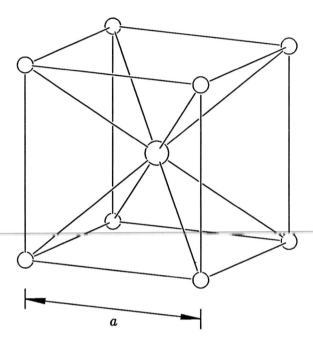

Fig. 1-25. The conventional unit cell of the cesium chloride structure. The large and small spheres represent two different types of atoms.

Problem 1-16. Using data in Appendix 3, find the volume of a primitive unit cell in a crystal of cesium chloride (CsCl). Answer: 69.4 Å^3.

1-11 Zincblende Structure

Next, let us examine the structure of ZnS (zinc sulfide or "zincblende"). This structure is more difficult to visualize than those which we already discussed. The conventional unit cell is shown in Fig. 1-26. The Zn atoms (large spheres) sit at fcc lattice points. If we imagine this unit cell to be divided into eight smaller cubes, then the S atoms (small spheres) sit at the center of four of these cubes, as shown in Fig. 1-26. Another way of describing the positions of the S atoms is to

CHAPTER 1 CRYSTAL STRUCTURE   27

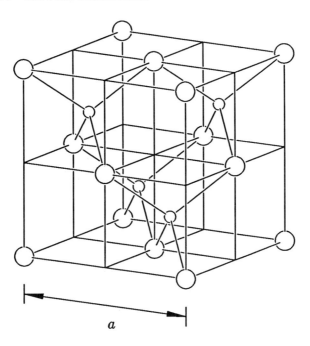

Fig. 1-26. The conventional unit cell for the zincblende structure. The large and small spheres represent two different types of atoms.

say that each S atom is displaced from a Zn atom by a vector $\frac{1}{4}a\hat{\imath} + \frac{1}{4}a\hat{\jmath} + \frac{1}{4}a\hat{k}$. The Bravais lattice of the ZnS structure is, of course, fcc. The conventional unit cell is the same as that of its fcc Bravais lattice and contains four lattice points (see Fig. 1-26). Thus the conventional unit cell contains four Zn atoms and four S atoms. The Wigner-Seitz primitive unit cell is that of its fcc Bravais lattice, the rhombic dodecahedron in Fig. 1-19, and, of course, contains one Zn atom and one S atom. Examples of crystals with the zincblende structure are given in Appendix 3.

Problem 1-17. Using data in Appendix 3, find the volume of a primitive unit cell in a crystal of gallium arsenide (GaAs). Answer: 44.9 Å^3.

1-12 Diamond Structure

The structure of diamond (carbon crystal) is almost identical to that of ZnS. If we change each Z atom to a C atom and each S atom to a C atom, we obtain the diamond structure. The conventional unit cell is shown in Fig. 1-27. You might expect here that since all the atoms are of the same type, all atoms are now at equivalent positions, and we have a new Bravais lattice. Such is *not* the case. The sites formerly occupied by Zn and S atoms are *not* equivalent, even though they are now both occupied by C atoms. The environment of each kind of site is not identical. Note that in the ZnS structure of Fig. 1-26, all "small" atoms have four nearest neighbors arranged like that of Fig. 1-28a, while all "large" atoms have four nearest neighbors arranged like that of Fig. 1-28b. These two arrangements are different. Thus, even though all atoms in the diamond structure are the same type, we can still tell the

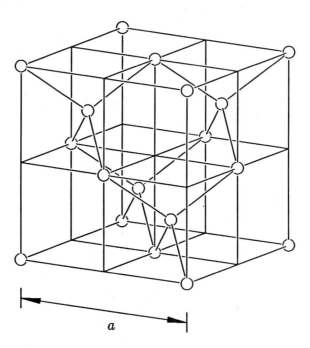

Fig. 1-27. The conventional unit cell of the diamond structure. All atoms are identical.

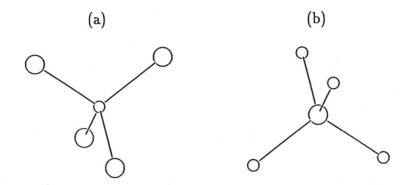

Fig. 1-28. Arrangement of nearest neighbors to (a) "small" atoms and (b) "large" atoms in the ZnS structure of Fig. 1-26.

difference between these two kinds of sites by looking at the arrangement of nearest neighbors. The Bravais lattice of diamond is therefore fcc, the same as for ZnS. The conventional unit cell contains eight C atoms, and the primitive unit cell contains *two* C atoms. Examples of crystals with the diamond structure are given in Appendix 3.

Problem 1-18. Using the data in Appendices 2 and 3, find the density of diamond (carbon crystal) in g/cm^3. Answer: 3.54 g/cm^3.

Problem 1-19. Using data in Appendix 3, find the volume of a primitive unit cell in a crystal of silicon (Si). Answer: 39.8 Å^3.

Problem 1-20. Using the data in Appendix 3, find the atomic diameter of carbon (C). Repeat for silicon (Si) and germanium (Ge). (Consider the atoms to be spheres that touch each other.) Answer: 1.54 Å, 2.35 Å, 2.44 Å.

1-13 Other Bravais Lattices

Since we were not successful in forming a new Bravais lattice with the diamond structure, we might wonder what other kinds of structures exist which have Bravais lattices different from those we already discussed. Actually, the restrictions of a Bravais lattice are quite severe. There are only fourteen mathematically possible Bravais lattices in all (see Fig. 1-29). Only three of these are cubic, i.e., the sc, bcc, and fcc lattices which we have discussed. The other eleven Bravais lattices are non-cubic.

In the triclinic lattice (Fig. 1-29a), all three basis vectors are unequal in length, and the angles between them are arbitrary.

In the primitive monoclinic lattice (Fig. 1-29b), all three basis vectors are also unequal in length, but *one* of them (the vertical one in the figure) is perpendicular to the other two. Just as we were able to produce the bcc and fcc lattices by adding extra lattice points to the sc lattice, we can produce the base-centered monoclinic lattice (Fig. 1-29c) by adding lattice points to *two* of the six faces of the primitive monoclinic lattice.

In the primitive orthorhombic lattice (Fig. 1-29d), all three basis vectors are perpendicular to each other, but they are still unequal in length. By adding extra points to this lattice, three more lattices can be produced: the base-centered orthorhombic, body-centered orthorhombic, and face-centered orthorhombic lattices (Figs. 1-29e-g).

In the primitive tetragonal lattice (Fig. 1-29h), all three basis vectors are perpendicular to each other, and two of them (the two in the x-y plane in the figure) are equal in length as well. By adding a point to this lattice, we can produce the body-centered tetragonal lattice (Fig. 1-29i).

In the hexagonal lattice (Fig. 1-29j), two of the basis vectors (the two in the x-y plane in the figure) are equal in length. The angle between them is 120°, and they are also both perpendicular to the third lattice vector. In the trigonal lattice (Fig. 1-29k), all three lattice vectors are equal in length and have the same angle with each other (but not 90°, otherwise

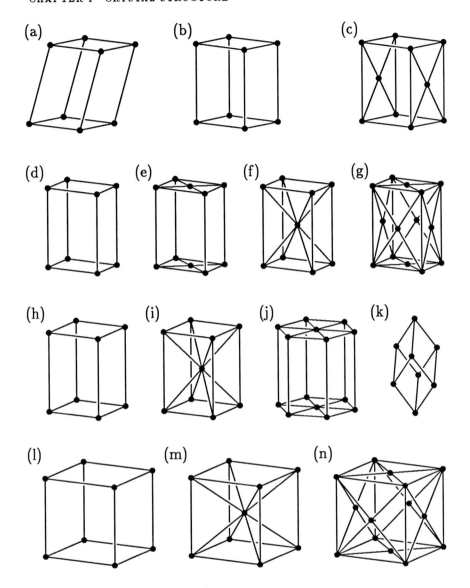

Fig. 1-29. The conventional unit cells of the 14 Bravais lattices. (a) triclinic, (b) primitive monoclinic, (c) base-centered monoclinic, (d) primitive orthorhombic, (e) base-centered orthorhombic, (f) body-centered orthorhombic, (g) face-centered orthorhombic, (h) primitive tetragonal, (i) body-centered tetragonal, (j) hexagonal, (k) trigonal, (l) primitive cubic (sc), (m) body-centered cubic, and (n) face-centered cubic.

it would be cubic). We will not discuss the non-cubic lattices further in this book.

1-14 Ionic Bond

Let us now discuss the forces which hold atoms together in a crystal. These forces are called **bonds**. As an example, consider a simple case: the bond between a sodium ion (Na^+, a sodium atom with one missing electron) and a chloride ion (Cl^-, a chlorine atom with an extra electron). The Na^+ has a charge $+e$, and the Cl^- has a charge $-e$, where $e = 1.602 \times 10^{-19}$ C, the magnitude of the charge of an electron. The Na^+ and Cl^- are thus attracted to each other by an electrostatic force given by Coulomb's Law,

$$F = -\frac{e^2}{4\pi\epsilon_0 r^2}, \qquad (1\text{-}10)$$

where r is the separation of the two ions and ϵ_0 is the permittivity constant (see Appendix 1). The negative sign means that the force is in such a direction that it tries to *decrease r*. In other words, the force is *attractive*. Obviously, if this were the only force present, the two ions would move toward each other until $r = 0$. To prevent this from happening, a *repulsive* force must be present. Each ion consists of a positively charged nucleus surrounded by a "cloud" of electrons. When the two ions are very close to each other, their electron clouds repel each other. This repulsion is due to the electrostatic force that exists between electrons (like charges repel each other) and is also due to a quantum mechanical effect which is beyond the scope of our present discussion.

The repulsive force between the Na^+ and Cl^- ions is *short-ranged*. It is negligibly small except when the two ions are very close to each other, and then it is very large. This situation is very much like the forces between two "hard" spheres which are oppositely charged. An approximate plot of the repulsive force between a Na^+ and Cl^- ion is shown in Fig. 1-30, along with a plot of the attractive force for comparison. The net force is given by the sum of the attractive and repulsive forces

CHAPTER 1 CRYSTAL STRUCTURE   33

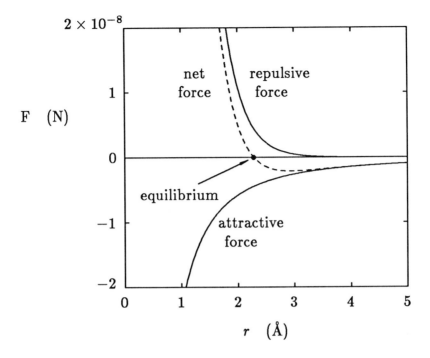

Fig. 1-30. The repulsive, attractive, and net force between a Na$^+$ and a Cl$^-$ ion separated by distance r.

and is shown as a dashed line in the figure. At large r the net force is attractive, and at small r the net force is repulsive. A point of stable equilibrium exists where the net force is zero, as shown in the figure.

Problem 1-21. The repulsive force between Na$^+$ and Cl$^-$ ions is often written as

$$F = Be^{-r/R},$$

where $B = 5.45 \times 10^{-6}$ N and $R = 0.321$ Å. At equilibrium, this force is equal to the magnitude of the attractive force given by Eq. (1-10). Find the equilibrium value of r. Answer: 2.28 Å.

Problem 1-22. Consider an isolated Na$^+$-Cl$^-$ pair at equilibrium separation, $r = 2.28$ Å (see Problem 1-21). Calculate the work required to separate the two ions by an infinite distance. This work is called the binding energy of the bond. Answer: 5.42 eV.

In the formation of a crystal of NaCl, each Na atom gives up an electron to a Cl atom so that the crystal consists of Na^+ and Cl^- ions. Each Na^+ has six nearest-neighbor Cl^- ions, and each Cl^- has six nearest-neighbor Na^+ ions. Thus, there is a large attractive force present which holds the crystal together. The short-range repulsive force between nearest neighbors tends to push ions away from each other when they get too close. There exists a position of equilibrium for the crystal when the lattice constant is $a = 5.63$ Å for NaCl. For a smaller lattice constant, $a < 5.63$ Å, the repulsive forces are stronger than the attractive forces, and the crystal tries to expand. For a larger lattice constant, $a > 5.63$ Å, the attractive forces are stronger than the repulsive forces, and the crystal tries to contract. At $a = 5.63$ Å, the forces are in equilibrium. The type of bonding force which exists between the Na^+ and Cl^- ions in NaCl is called an **ionic bond**. Many crystals are held together by ionic bonds.

Problem 1-23. The equilibrium distance between an isolated Na^+-Cl^- pair is 2.28 Å (see Problem 1-21). In a NaCl crystal, the equilibrium distance between Na^+-Cl^- pairs of nearest neighbors is 2.82 Å (see Problem 1-11). Why is this distance greater in a crystal than in an isolated pair?

1-15 Covalent Bond

In a crystal like diamond, no ions are present. The crystal consists entirely of electrically neutral carbon (C) atoms. The type of bond which holds diamond together is called a **covalent bond**. Each pair of nearest-neighbor C atoms forms a bond by *sharing* their electrons. A covalent bond between two C atoms in diamond consists of two electrons, one from each atom. These two electrons spend most of their time *between* the two atoms. Since each atom lost an electron to the bond, each atom is positively charged and each is attracted to the pair of electrons between them. This results in a net attractive force between the two atoms. At equilibrium, this attractive force is balanced by a repulsive force of the same type as in an ionic bond. (Actually, an accurate description of covalent bonds is quantum mechanical and is beyond the scope of our present discussion.) In the diamond structure, each atom has four nearest neighbors (see Fig. 1-27). Thus, each C atom gives up four electrons to form the covalent bonds to its neighboring C atoms. Other crystals with the diamond structure are also held together by covalent bonds.

Problem 1-24. Consider a diamond crystal containing 1 mole of carbon (C) atoms. (1 mole is equal to Avogadro's number. See Appendix 1.) It is found that the energy required to separate all the C atoms from each other to an infinite distance apart is 700 kJ (called the cohesive energy of the crystal). Find the binding energy (see Problem 1-22) of a single C-C covalent bond. Answer: 3.6 eV.

1-16 Metallic Bond

In metals, the atoms are held together by a different kind of bond which is neither ionic nor covalent. It is called the **metallic bond**. In a metal like sodium (Na), for example, each Na contributes an electron to the bond. These electrons are shared by *all* the atoms. In a covalent bond, each electron is shared by *two* atoms. In the metallic bond, each electron is shared by *all* the atoms. These electrons are free to roam about anywhere in the crystal. In an electric field, these electrons can form an electric current. Thus we call them **conduction electrons**.

The attractive force between atoms in a metallic bond arises from two sources. First of all, like the covalent bond, the electrons spend most of their time *between* the positively charged Na atoms. The Na atoms are attracted to these electrons and thus are effectively attracted to each other. Secondly, the kinetic energy of these electrons is much lower when they are free to roam about in the crystal than when they are bound tightly to the atoms. The conduction electrons in the Na metal are thus in a state of lower energy than they would be in a simple collection of free atoms. We will discuss the nature of the conduction electrons in more detail in later chapters.

CHAPTER 2

X-RAY DIFFRACTION

2-1 Waves

X-ray diffraction is the method most commonly used to determine the structure of a crystal. As an introduction, let us first discuss wave phenomena in general. A traveling sine wave is given by the equation,

$$y = A\sin(kx - \omega t). \qquad (2\text{-}1)$$

This may represent, for example, a wave traveling down a long horizontal rope, where t is the time, x is the distance down the rope, and y is the vertical displacement of the rope at position x and time t. The quantity A is called the **amplitude** of the wave. If we "freeze" the wave at $t = 0$, it would look like Fig. 2-1. As a function of time, this entire wave form moves to the right with some velocity v.

The **wavelength** λ of this wave is defined to be the distance between adjacent maxima in y at any given instant of time, as shown in Fig. 2-1. During one wavelength, the sine function in Eq. (2-1) goes through one cycle, which is 2π radians. Thus, when x changes by an amount λ, the quantity kx

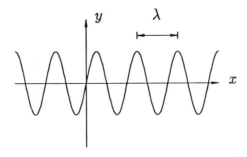

Fig. 2-1. Traveling sine wave given by Eq. (2-1) at $t = 0$.

must change by an amount 2π. This requires $k\lambda = 2\pi$, or

$$k = 2\pi/\lambda, \qquad (2\text{-}2)$$

where k is called the **wave number** and has dimensions of inverse length (m^{-1} in SI units).

The **frequency** ν of the wave is the rate at which maxima pass by a fixed point on the rope. The units of frequency are inverse seconds (s^{-1}) or hertz (Hz). The **period** T of the wave is the time required for the wave to travel one wavelength and is simply related to the frequency ν by

$$T = 1/\nu. \qquad (2\text{-}3)$$

During the time required for the wave to travel one wavelength, the sine function in Eq. (2-1) goes through one cycle which is 2π radians. Thus, when t changes by an amount T, the quantity ωt must change by an amount 2π. This requires $\omega T = 2\pi$, or, using Eq. (2-3),

$$\omega = 2\pi\nu, \qquad (2\text{-}4)$$

where ω is called the **angular frequency** and has units of rad/s or just s^{-1} since radians are dimensionless units. (Angular frequency is *never* expressed in units of hertz.)

Since the wave travels one wavelength in one period, we can obtain the velocity $v = \lambda/T$, or, using Eq. (2-3),

$$v = \lambda\nu, \qquad (2\text{-}5)$$

or, combining Eqs. (2-2), (2-4), and (2-5),

$$\omega = vk. \qquad (2\text{-}6)$$

Problem 2-1. Consider a sound wave of frequency $\nu = 262$ Hz ("middle C" in music). The velocity of sound is $v = 340$ m/s. Find the wavelength λ, the wave number k, the period T, and the angular frequency ω of the wave. Answer: 1.30 m, 4.84 m^{-1}, 3.82 ms, 1650 s^{-1}.

Problem 2-2. Yellow light has a wavelength of about 550 nm. The velocity of light is $c = 3.00 \times 10^8$ m/s. Find the frequency ν, the angular frequency ω, the period T, and the wave number k of the wave. Answer: 5.45×10^{14} Hz, 3.43×10^{15} s^{-1}, 1.83×10^{-15} s, 1.14×10^7 m^{-1}.

2-2 Interference

When more than one wave is present, we may observe interference phenomena. As an example, consider a pair of speakers which emit a sound wave of wavelength λ. We drive each speaker with a common amplifier so that a maximum of each wave is emitted from the speakers simultaneously. (The speakers are said to be "in phase.") If we place these speakers side by side, the two waves will travel on top of each other, through the same region of space (see Fig. 2-2). Under most circumstances, such waves act independent of each other. Thus, we can obtain the resultant wave by algebraic addition of the two waves. In Fig. 2-2, the resultant wave is simply a wave of twice the amplitude of either of the two waves. This is a case of **constructive interference** between two waves.

In contrast, consider what happens if we move back one of the speakers by a half wavelength (see Fig. 2-3). We can see

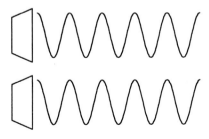

Fig. 2-2. Sound waves from a pair of speakers. These waves constructively interfere.

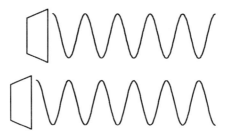

Fig. 2-3. Sound waves from a pair of speakers. One speaker is moved back by a half wavelength. These waves destructively interfere.

that when one wave is up, the other is down, and vice versa. The algebraic sum of these two waves is zero. The resultant wave has zero amplitude. The two waves have canceled each other, and we will hear no sound. This is a case of **destructive interference** between two waves.

If we now move back the speaker *another* half wavelength so that it sits one whole wavelength behind the other speaker, the two waves again constructively interfere (see Fig. 2-4). In general, the two waves will constructively interfere if the distance Δx between the two speakers is a multiple of wavelengths.

If we place the speakers not side by side, as in the example

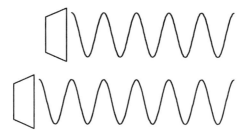

Fig. 2-4. Sound waves from a pair of speakers. One speaker is moved back by a whole wavelength. These waves constructively interfere.

above, but each at some arbitrary position relative to a listener, we can examine a more general situation (see Fig. 2-5). Let x_1 be the distance between speaker #1 and the listener, and let x_2 be the distance between speaker #2 and the listener. Certainly, if $x_1 = x_2$, then the maxima in the wave from each speaker will arrive at the listener at the same time and the two waves will constructively interfere. This will also occur if one speaker is any multiple of wavelengths closer to the listener than the other. Thus, the general condition for constructive interference is

$$x_2 - x_1 = n\lambda, \tag{2-7}$$

where n is any integer (including negative integers and also $n = 0$).

On the other hand, if one speaker is a *half* wavelength closer to the listener than the other speaker, the two waves will destructively interfere. This will also occur if the speaker is closer by $\frac{3}{2}\lambda$, $\frac{5}{2}\lambda$, $\frac{7}{2}\lambda$, etc. Thus, the general condition for destructive interference is

$$x_2 - x_1 = (n + \tfrac{1}{2})\lambda. \tag{2-8}$$

Of course, destructive interference will be complete (i.e., *zero* resultant amplitude) *only* if the amplitude of each of the two waves are exactly equal at the listener.

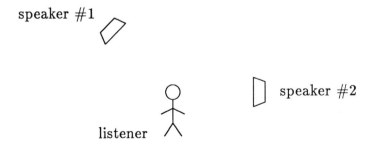

Fig. 2-5. A person listening to two speakers.

Problem 2-3. Two speakers are located 4.0 m apart on the stage of an auditorium. A listener is seated 25.0 m from one speaker and 27.0 m from the other. We drive the two speakers at a frequency which we sweep from 100 Hz to 500 Hz. If the two speakers are in phase, find the frequencies at which the waves constructively interfere at the listener so that he hears maximum intensity. Find the frequencies at which the waves destructively interfere at the listener so that he hears minimum intensity. The speed of sound is 340 m/s. Answer: 170 Hz, 340 Hz, 255 Hz, 425 Hz.

2-3 Multiple-Slit Diffraction

Another example of wave interference is the multiple-slit diffraction of light waves. Consider a "diffraction grating" consisting of a large number of parallel slits separated by distance d. We shine monochromatic light (from a laser, for example) through the grating and observe the pattern of light on a distant screen (far away compared to the dimensions of the grating) as shown in Fig. 2-6. Each illuminated slit in the grating behaves like an individual source which sends light waves out in all directions. Thus a grating with N slits across the laser beam produces, in effect, N individual sources of light, all in phase with each other since they each come from a common source.

Consider some point on the screen which is at an angle θ from the original direction of the laser beam. Each of the N "sources" contributes to the intensity of light at that point. Now, if the screen is "distant," the direction from each slit to that point on the screen will be very nearly parallel. Consider a pair of adjacent slits. As the two light waves emerge from the slits at an angle θ, one of them gets behind the other by an amount Δx (see Fig. 2-7). The two waves will constructively interfere only if Δx is some multiple number of wavelengths, i.e., $\Delta x = n\lambda$, where n is any integer. If we examine the

CHAPTER 2 X-RAY DIFFRACTION

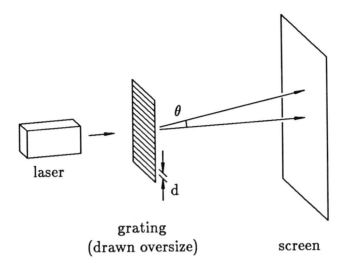

Fig. 2-6. Multiple-slit diffraction.

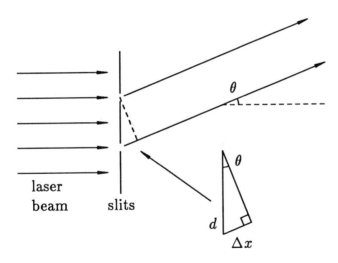

Fig. 2-7. Interference of light waves emerging from a pair of adjacent slits in a diffraction grating.

triangle shown in Fig. 2-7, we see that $\Delta x = d\sin\theta$, so that we finally obtain

$$d\sin\theta = n\lambda. \qquad (2\text{-}9)$$

When that equation is satisfied, the light waves from all pairs of adjacent slits constructively interfere, and therefore the light waves from all the slits across the laser beam will constructively interfere with each other, causing a rather intense spot of light to appear on the screen. For example, if $d = 2.0$ μm and $\lambda = 633$ nm, then, from Eq. (2-9), we obtain $\theta = 0, \pm 18°, \pm 39°, \pm 72°$ for $n = 0, \pm 1, \pm 2, \pm 3$, respectively. Thus we would observe a central spot at $\theta = 0$ (along the original direction of the laser beam), and three "diffraction" spots both above and below the central spot.

Problem 2-4. Consider a beam of laser light (wavelength $\lambda = 633$ nm) aimed at a screen. If we place a diffraction grating into the path of the laser beam, new spots appear on the screen, one above and one below the original spot. If the new spots are 0.40 m from the original spot and if the distance between the grating and the screen is 1.00 m, find the distance between the slits in the grating. Answer: 1.70 μm.

2-4 X-Ray Diffraction in Crystals

In the case of multiple-slit diffraction, we can see how interference phenomena can be used to determine the distance between slits in a diffraction grating. We can use a similar approach for determining the distance between atoms in a crystal. If we shine electromagnetic radiation into a crystal, each atom in the crystal individually scatters some of the radiation. Thus, a crystal with N atoms becomes N different sources of radiation. The waves radiated from these N sources interfere with each other, and, as with the diffraction grating, we find at some angles that these waves constructively interfere, giving rise to intense spots on a screen.

CHAPTER 2 X-RAY DIFFRACTION

In general, interference effects of waves can usually be observed in any object which contains a number of repeating units (slits in a grating, atoms in a crystal, etc.). For the effect to be most easily measurable, we must use waves with a wavelength comparable to the distance between the repeating units. As an example, for a diffraction grating, we need to use $\lambda \cong d$. If we use $\lambda > d$, there is no solution to Eq. (2-9) except for $n = 0$ so that we only observe the central spot at $\theta = 0$. We observe *no* diffraction spots. If we use $\lambda \ll d$, then we see from Eq. (2-9) that the diffraction spots are very close together. Overlap of the spots may cause them to be unresolvable.

Similarly, to obtain measurable interference effects in crystals, we need to use waves with a wavelength comparable to the distance between atoms, which is a few angstroms. Electromagnetic radiation with that wavelength is called **x rays**. The phenomenon of interference of these waves by a crystal is called **x-ray diffraction**. The x rays most commonly used for x-ray diffraction have a wavelength $\lambda = 1.542$ Å.

Let us now take a closer look at the details of x-ray diffraction in a crystal. First, consider the scattering of waves from a single *plane* of atoms (see Fig. 2-8). If $\theta' = \theta$, then the scattered waves will constructively interfere (see Problem 2-5).

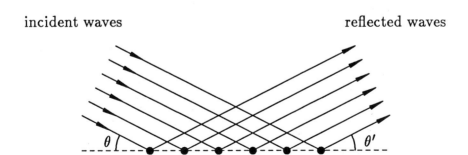

Fig. 2-8. Reflection of waves from a plane of atoms.

Problem 2-5. Prove that if $\theta' = \theta$ in Fig. 2-8, then the scattered waves will be in phase with each other and thus constructively interfere. Assume that the incident waves are in phase with each other.

In a crystal, parallel planes of atoms are equally spaced. As shown above, each of these planes reflects the incident waves at an angle $\theta' = \theta$. However, since reflected waves from different planes are superimposed on each other, we must also examine the interference between *these* waves. Consider a pair of adjacent planes separated by a distance d (see Fig. 2-9). From the figure, we see that waves reflected from the lower plane get behind the waves reflected from the upper plane. The amount by which the lower wave gets behind is given by the distance it travels from point A to point B in the figure. From the triangle in the figure, we see that this distance is $2d \sin \theta$. If we want the interference between the two waves to be constructive, then the amount by which the lower wave gets behind the upper wave must be some multiple number of

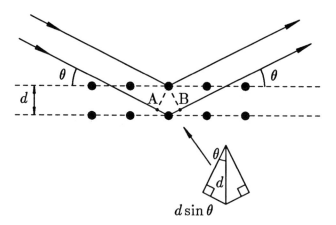

Fig. 2-9. Reflection of waves from two adjacent planes of atoms.

wavelengths. Thus,
$$2d\sin\theta = n\lambda, \qquad (2\text{-}10)$$
where n is an integer. This equation is called **Bragg's Law**. The angles θ which satisfy this equation are called **Bragg angles**. If waves reflected from two adjacent planes of atoms constructively interfere, then the waves reflected from all parallel planes of atoms in the crystal also constructively interfere, since the planes are evenly spaced. Thus, Bragg's Law gives the direction at which we should be able to observe reflection of x rays from the crystal.

As an example, consider diffraction of x rays ($\lambda = 1.542$ Å) from planes of atoms separated by a distance $d = 3.0$ Å. From Eq. (2-10), we find that $\sin\theta = n\lambda/2d$, and thus $\theta = 15°, 31°, 51°$ for $n = 1, 2, 3$, respectively.

Problem 2-6. Consider a set of crystal planes which are separated by 1.95 Å. If we use x rays of wavelength $\lambda = 1.542$ Å, find all possible Bragg angles for reflection from these planes. Answer: 23°, 52°.

Problem 2-7. We observe a Bragg angle $\theta = 15.3°$ for reflection of x rays of wavelength $\lambda = 1.542$ Å from some set of planes in a crystal. Find the distance between these planes. (Consider $n = 1$ in Bragg's Law.) Answer: 2.92 Å.

Problem 2-8. We observe reflection of x rays at the Bragg angle $\theta = 18°$. If this is a "first-order" Bragg reflection [i.e., $n = 1$ in Eq. (2-10)], find the Bragg angles for all of the higher-order reflections ($n = 2, 3$, etc.). You do not need to know the wavelength of the x rays or the distance between crystal planes to solve this problem. Answer: 38°, 68°.

Problem 2-9. When copper is heated from 0°C to 100°C, its lattice parameter a increases by 0.17% due to thermal expansion. If we observe an x-ray reflection at a Bragg angle $\theta = 19.3°$ at 0°C, by how much will θ change when the sample is heated to 100°C? (You do not need to know λ or d to solve this problem.) Answer: $-0.034°$.

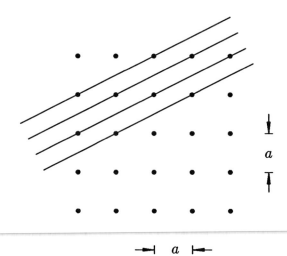

Fig. 2-10. Figure for Problem 2-10.

Problem 2-10. Consider planes shown for the two-dimensional square lattice in Fig. 2-10. If $a = 4.00$ Å and we use x rays of wavelength 1.542 Å, at what Bragg angles would we observe reflections from these planes? Answer: 26°, 60°.

As an example using a real crystal, consider x-ray diffraction in copper (Cu). Cu has an fcc lattice with $a = 3.61$ Å (see Appendix 3). In Problem 1-4, we found that the (100) planes in Cu are separated by a distance $d = 1.81$ Å. Using x rays of wavelength $\lambda = 1.542$ Å, we obtain from Eq. (2-10) the Bragg angles $\theta = 25°$ and $58°$ for $n = 1$ and 2, respectively. We would detect reflection of the x rays from the (100) planes at each of these angles. Note that the crystal must be oriented so that the incident angle of the x rays is equal to θ as well.

If we orient the Cu crystal correctly, we can also detect reflection from other sets of planes as well. For example, from Problem 1-4, we found that the (110) planes in Cu are separated by a distance $d = 1.27$ Å. We would be able to detect reflection of x rays from these planes at the angle $\theta = 37°$ ($n = 1$).

Problem 2-11. Find all Bragg angles less than 40° for reflection of x rays ($\lambda = 1.542$ Å) from the (100) and (110) planes in sodium (Na). Answer: 21°, 15°, 30°.

2-5 Diffraction from Real Atoms

Up to now, we have been considering x rays to be scattered by "points" in the crystal (see Fig. 2-9, for example). Actually, atoms are not "points" but have spatial dimensions. Each part of the atom scatters x rays differently. To see how Bragg's Law applies to this more realistic picture, we must use the concept of primitive unit cell discussed in Chapter 1. Since every primitive unit cell in the crystal is identical in contents, they each scatter x rays in an identical manner.

Consider some point in the primitive unit cell. There is an equivalent point in each of the other cells in the crystal. This set of equivalent points forms a Bravais lattice of the crystal. (Remember that the position of the lattice points with respect to the unit cells is arbitrary.) If we consider the scattering of x rays from just this set of points (see Fig. 2-11), we again find ourselves in a situation similar to that which led us to Bragg's law. These points form planes, and the x rays scattered from these planes will constructively interfere only if Eq. (2-10) is satisfied. In this case, we apply Bragg's Law to a set of equivalent points in the crystal rather than to positions of atoms themselves.

We can repeat this procedure for other points in the primitive unit cell as well. For each point, there is an equivalent point in each of the other primitive unit cells in the crystal, and these equivalent points always form the same Bravais lattice. Thus, waves are scattered from each set of equivalent points at the *same* Bragg angles.

The *total* scattered wave, which is a superposition of the waves scattered from each set of equivalent points, can therefore only have intensity at these Bragg angles. Of course, waves scattered from different points in the primitive unit cell also

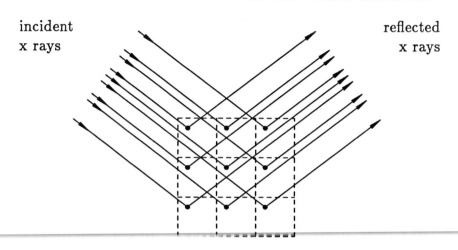

Fig. 2-11. Scattering of x rays from equivalent points, one in each primitive unit cell. The points form the Bravais lattice of the crystal.

interfere with each other. This interference affects the intensity of the total scattered wave at the Bragg angle. In fact, it is possible to obtain total destructive interference among these waves, so that although Bragg's Law may *allow* a certain reflection to be present, we may not observe it because of interference *within* the primitive unit cell.

Generally, then, Bragg's Law must be applied to points of the Bravais lattice of a crystal rather than to positions of atoms. The distance d in Eq. (2-10) refers to the distance between planes of Bravais *lattice* points, rather than to planes of atoms. In our example of Cu, the atomic positions and the Bravais lattice are the same, so that there is no difference between planes of Bravais lattice points and planes of atoms. However, in a crystal like cesium chloride (CsCl), the difference is important.

The Bravais lattice of CsCl is sc with $a = 4.11$ Å (see Appendix 3). The (100) planes of the sc lattice points are separated by a distance $d = a = 4.11$ Å. Thus for x rays of wavelength $\lambda = 1.542$ Å, the Bragg angles are $\theta = 11°, 22°, 34°, 47°, 70°$ for $n = 1, 2, 3, 4, 5$, respectively. Notice that (100) planes of *atoms* in CsCl are separated by a distance $d = \frac{1}{2}a = 2.06$ Å (see Fig. 1-25). If we had used this distance in Bragg's Law, we would have obtained the Bragg angles $\theta = 22°$ and $47°$. Thus, the Bragg reflections observed at $\theta = 11°$, $34°$, and $70°$ would not have been predicted.

Problem 2-12. What is the Bravais lattice of sodium chloride (NaCl)? Using x rays of wavelength $\lambda = 1.542$ Å, find all the Bragg reflections from the (100) planes of NaCl. Repeat for the (110) planes. Answer: $16°$, $33°$, $55°$, $23°$, $51°$.

2-6 Reciprocal Lattice

Thus far, we have only considered Bragg reflection from (100) and (110) planes. There are, of course, many other planes in a crystal which can give rise to Bragg reflections as well. To find all the possible Bragg reflections in a crystal is a rather difficult task using Bragg's Law in the form of Eq. (2-10). Let us now find another form of Bragg's Law which is easier to apply to a general reflection.

We want to examine the interference of waves scattered by a set of Bravais lattice points. Consider two lattice points separated by a vector **R**. An incoming x ray is scattered at these two points in all directions. We will consider some particular direction of the outgoing x ray. Let θ_1 and θ_2 be the angles between **R** and the incoming and outgoing x rays, respectively (see Fig. 2-12). After being scattered at the lattice points, one outgoing x ray will be behind the other by an amount

$$\Delta x = R\cos\theta_1 + R\cos\theta_2. \qquad (2\text{-}11)$$

The two x rays will constructively interfere only if Δx is a multiple number of wavelengths, i.e., $\Delta x = n\lambda$, where λ is the wavelength of the x rays. Combining this with Eq. (2-11), we obtain

$$R\cos\theta_1 + R\cos\theta_2 = n\lambda. \qquad (2\text{-}12)$$

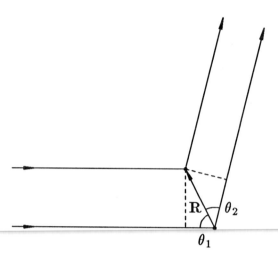

Fig. 2-12. X rays scattered by two lattice points.

Now let us represent a wave by a **wave vector k** which points along the direction of the wave. The magnitude of **k** is the wave number $k = 2\pi/\lambda$. Let $\mathbf{k}_1$ and $\mathbf{k}_2$ be wave vectors for the incoming and outgoing x rays, respectively. From Fig. 2-13, we see that

$$\mathbf{R} \cdot \mathbf{k}_1 = Rk\cos(\pi - \theta_1) = -Rk\cos\theta_1 \qquad (2\text{-}13)$$

and

$$\mathbf{R} \cdot \mathbf{k}_2 = Rk\cos\theta_2. \qquad (2\text{-}14)$$

Combining Eqs. (2-12)–(2-14), we obtain

$$\mathbf{R} \cdot (\mathbf{k}_2 - \mathbf{k}_1) = nk\lambda = 2\pi n, \qquad (2\text{-}15)$$

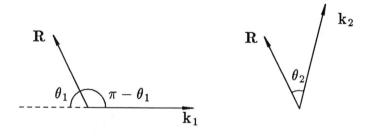

Fig. 2-13. Wave vectors, $\mathbf{k}_1$ and $\mathbf{k}_2$, of incoming and outgoing x rays, respectively.

CHAPTER 2 X-RAY DIFFRACTION

or
$$\mathbf{R} \cdot \mathbf{G} = 2\pi n, \qquad (2\text{-}16)$$

where
$$\mathbf{G} = \mathbf{k}_2 - \mathbf{k}_1. \qquad (2\text{-}17)$$

In order for *all* outgoing x rays in the crystal to constructively interfere with each other, we must find a vector $\mathbf{G}$ which satisfies Eq. (2-16) for *all* vectors $\mathbf{R}$ between pairs of lattice points in the crystal. These vectors $\mathbf{R}$ are lattice vectors of the Bravais lattice and can be written in the form,

$$\mathbf{R} = n_1 \mathbf{a}_1 + n_2 \mathbf{a}_2 + n_3 \mathbf{a}_3, \qquad (2\text{-}18)$$

where $\mathbf{a}_1, \mathbf{a}_2, \mathbf{a}_3$ are basis vectors of the Bravais lattice and n_1, n_2, n_3 are integers. Obviously, Eq. (2-16) will be satisfied for any $\mathbf{G}$ which satisfies the following three equations:

$$\begin{aligned} \mathbf{a}_1 \cdot \mathbf{G} &= 2\pi m_1, \\ \mathbf{a}_2 \cdot \mathbf{G} &= 2\pi m_2, \\ \mathbf{a}_3 \cdot \mathbf{G} &= 2\pi m_3, \end{aligned} \qquad (2\text{-}19)$$

where m_1, m_2, m_3 are integers.

To show how this is done, let us consider an example: the sc lattice. The basis vectors are given by Eq. (1-5),

$$\begin{aligned} \mathbf{a}_1 &= a\hat{\imath}, \\ \mathbf{a}_2 &= a\hat{\jmath}, \\ \mathbf{a}_3 &= a\hat{k}. \end{aligned} \qquad (2\text{-}20)$$

Eqs. (2-19) then become

$$\begin{aligned} G_x &= m_1(2\pi/a), \\ G_y &= m_2(2\pi/a), \\ G_z &= m_3(2\pi/a), \end{aligned} \qquad (2\text{-}21)$$

and we have

$$\mathbf{G} = m_1 \frac{2\pi}{a}\hat{\imath} + m_2 \frac{2\pi}{a}\hat{\jmath} + m_3 \frac{2\pi}{a}\hat{k}. \qquad (2\text{-}22)$$

Remember that m_1, m_2, m_3 are arbitrary integers. Any choice of m_1, m_2, m_3 will give us a $\mathbf{G}$ which satisfies Eq. (2-16). There are an infinite number of different vectors $\mathbf{G}$ which satisfy Eq. (2-16).

Written in the form of Eq. (2-22), $\mathbf{G}$ looks very much like some kind of a lattice vector [compare with Eq. (2-18)]. If we define basis vectors

$$\mathbf{b}_1 = (2\pi/a)\hat{\imath},$$
$$\mathbf{b}_2 = (2\pi/a)\hat{\jmath}, \qquad (2\text{-}23)$$
$$\mathbf{b}_3 = (2\pi/a)\hat{k},$$

then

$$\mathbf{G} = m_1 \mathbf{b}_1 + m_2 \mathbf{b}_2 + m_3 \mathbf{b}_3. \qquad (2\text{-}24)$$

The vectors $\mathbf{G}$ give us a set of points which form a Bravais lattice. In this case, the lattice is sc with lattice parameter $2\pi/a$, as can be seen in Eq. (2-23). The dimensions of "distance" in this lattice are inverse length, the same units as the wave number k. The space in which this lattice exists is called "reciprocal space" or "k-space" (in contrast to "real space" where lattices described by $\mathbf{R}$ exist). The lattice in reciprocal space is called the **reciprocal lattice**. The original lattice in real space is called the **direct lattice**.

For a general Bravais lattice with lattice vectors $\mathbf{R}$ given in the form of Eq. (2-18), the solution to Eq. (2-16) can *always* be written in the form of Eq. (2-24). In other words, $\mathbf{G}$ is a vector connecting points of a reciprocal lattice. For every direct lattice in real space, there exists a corresponding reciprocal lattice in reciprocal space. The basis vectors $\mathbf{b}_1, \mathbf{b}_2, \mathbf{b}_3$ of the reciprocal lattice in reciprocal space can be written in terms of the basis vectors a_1, a_2, a_3 of the direct lattice in real space as

$$b_1 = \frac{2\pi a_2 \times a_3}{a_1 \cdot (a_2 \times a_3)},$$

$$b_2 = \frac{2\pi a_3 \times a_1}{a_2 \cdot (a_3 \times a_1)}, \qquad (2\text{-}25)$$

$$b_3 = \frac{2\pi a_1 \times a_2}{a_3 \cdot (a_1 \times a_2)}.$$

Problem 2-13. Show that Eq. (2-24) with basis vectors given by Eq. (2-25) is a solution to Eq. (2-16). Do not assume a_1, a_2, a_3 are perpendicular to each other.

We already showed that the reciprocal lattice of an sc lattice with lattice parameter a is an sc lattice with lattice parameter $2\pi/a$. From Eq. (2-25), we can also show that the reciprocal lattice of a bcc lattice with lattice parameter a is an fcc lattice with lattice parameter $4\pi/a$. Also, we can show that the reciprocal lattice of an fcc lattice with lattice parameter a is a bcc lattice with lattice parameter $4\pi/a$. The reciprocal lattice is very useful in solid state physics, and we will refer to it again many times in this course.

Problem 2-14. Show that the reciprocal lattice of a bcc lattice with lattice parameter a is an fcc lattice with lattice parameter $4\pi/a$. Use Eq. (2-25).

Problem 2-15. Show that the reciprocal lattice of an fcc lattice with lattice parameter a is a bcc lattice with lattice parameter $4\pi/a$. Use Eq. (2-25).

Let us now return to the original problem of x-ray diffraction. From Eq. (2-17), remembering that the magnitudes of k_1

and k_2 are equal, we see that **G** must bisect the angle between k_1 and k_2 (see Fig. 2-14a). The direction of **G** is always perpendicular to some plane of lattice points in real space. (For cubic lattices, this property of the reciprocal lattice vector is obviously true. But it can also be shown to be true for noncubic lattices as well.) Drawing a plane perpendicular to **G** in Fig. 2-14b, we see that the outgoing wave (given by k_2) is a reflection of the incident wave (given by k_1) from this plane. The angle θ is defined the same way as in Bragg's Law, and we obtain from the triangle in Fig. 2-14c, remembering that the magnitudes of k_1 and k_2 are both equal to k,

$$G = 2k \sin \theta. \qquad (2\text{-}26)$$

This equation is equivalent to Bragg's Law, as we will now demonstrate by example.

Consider again the copper (Cu) crystal. The Bravais lattice in real space (the direct lattice) is fcc with lattice parameter $a = 3.61$ Å. The reciprocal lattice is bcc with lattice parameter $4\pi/a = 3.48$ Å^{-1}. Bragg reflections occur at angles

$$\sin \theta = G/2k. \qquad (2\text{-}27)$$

If we use x rays of wavelength $\lambda = 1.542$ Å, we find $k = 2\pi/\lambda = 4.07$ Å^{-1}. Eq. (2-27) can only have solutions for $\sin \theta \leq 1$, so

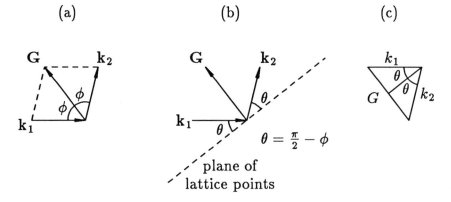

Fig. 2-14. (a) Diagram showing the relation, $\mathbf{G} = k_2 - k_1$. (b) **G** is perpendicular to planes of lattice points in real space. (c) Diagram showing the relation between G, k, and θ.

CHAPTER 2 X-RAY DIFFRACTION   57

Table 2-1. Bragg reflections from copper using x rays of wavelength 1.542 Å.

G	G	θ	planes
$(4\pi/a)\hat{i}$	3.48 Å^{-1}	25°	(100)
$(4\pi/a)2\hat{i}$	6.96	58	(100)
$(4\pi/a)(\hat{i}+\hat{j})$	4.92	37	(110)
$(4\pi/a)(\frac{1}{2}\hat{i}+\frac{1}{2}\hat{j}+\frac{1}{2}\hat{k})$	3.01	22	(111)
$(4\pi/a)(\hat{i}+\hat{j}+\hat{k})$	6.03	48	(111)
$(4\pi/a)(2\hat{i}+\hat{j})$	7.78	73	(210)
$(4\pi/a)(\frac{3}{2}\hat{i}+\frac{1}{2}\hat{j}+\frac{1}{2}\hat{k})$	5.77	45	(311)
$(4\pi/a)(\frac{3}{2}\hat{i}+\frac{3}{2}\hat{j}+\frac{1}{2}\hat{k})$	7.58	69	(331)

we must have $G \leq 2k = 8.14$ Å^{-1}. In Table 2-1, we list some of the G-vectors which satisfy this condition. Also listed for each vector is the corresponding length G of the vector, the Bragg angle θ calculated from Eq. (2-27), and the plane of reflection which is perpendicular to the direction of **G**. We recognize the Bragg angles for reflections from the (100) and (110) planes which we previously calculated using Bragg's Law in the form of Eq. (2-10). The other Bragg angles listed in the table could also be obtained using Eq. (2-10), but they are much more easily obtained using Eq. (2-27), as we did.

2-7 Experimental Methods

Using a single crystal of Cu, each of these Bragg reflections can be observed by correctly orienting the crystal with respect to the incident x ray. However, these reflections can only be observed one at a time since the crystal must be oriented in a different direction in order to observe each reflection. If we use a powdered sample of Cu, we can observe *all* Bragg reflections simultaneously. A powdered sample actually consists of many small crystals of Cu. Any one of those crystals can diffract the x ray if its orientation is correct. In a powdered sample con-

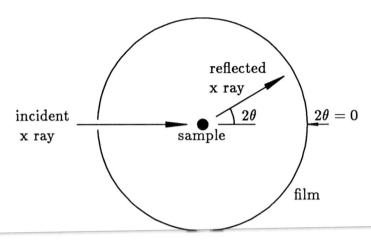

Fig. 2-15. X-ray diffraction from a powdered sample.

taining many small crystals randomly oriented, many of the orientations possible for a single crystal are represented simultaneously. Thus, for each of the Bragg reflections listed in Table 2-1, there are very likely a number of crystals in the sample which are oriented correctly for that particular reflection.

In practice, we encircle the sample with a long strip of photographic film (see Fig. 2-15). The reflected x rays expose the film, and we thus obtain a set of "diffraction lines," one for each of the possible Bragg angles in the sample. From Fig. 2-14b, we see that the direction of the x ray changes by 2θ when it is reflected. Thus, the angle measured for diffraction lines on the film is 2θ. In Fig. 2-16, we show what such a strip

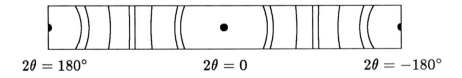

Fig. 2-16. X-ray diffraction lines on a strip of film. The sample is powdered copper. The wavelength of the x rays is 1.542 Å.

CHAPTER 2 X-RAY DIFFRACTION   59

of film would look like for Cu. There are eight lines which are those listed in Table 2-1.

Problem 2-16. Why are the diffraction lines on the strip of film in Fig. 2-16 curved?

Problem 2-17. Using a ruler, find the Bragg angle for each diffraction line shown in Fig. 2-16. Compare with Table 2-1 and identify each line by the planes of reflection.

Problem 2-18. An x-ray diffraction pattern from a powdered sample of lithium (Li) shows 10 lines. Find any 8 of them, giving the Bragg angle and the plane of Bragg reflection. The wavelength of the x rays is 1.542 Å. Answer: All 10 of them are 18°, 26°, 33°, 39°, 44°, 50°, 56°, 62°, 69°, 80°.

Problem 2-19. (a) Consider x-ray diffraction in a bcc lattice. Which planes will reflect the x rays at the smallest Bragg angle (call it θ_1)? Which planes will reflect the x rays at the next smallest Bragg angle (call it θ_2)? Find the ratio $\sin\theta_2/\sin\theta_1$. (b) Repeat part (a) for an fcc lattice. (c) Using x rays of wavelength $\lambda = 1.542$ Å, we find the reflections at the two smallest Bragg angles to be at 19.4° and 28.0°. Is the lattice bcc or fcc? Find the lattice parameter a. Answer: 1.41, 1.15, 3.28 Å.

Problem 2-20. Consider diffraction of x rays from sodium chloride (NaCl). (a) If we put NaCl under pressure, we find that its lattice parameter changes. At a pressure of about 300,000 atm, we find that the "first" diffraction line (the smallest Bragg angle) is at $\theta = 16.0°$, using x rays of wavelength 1.542 Å. Find the lattice parameter a of NaCl at that pressure. Find the density in g/cm^3. (b) If we now apply a bit more pressure, we find that the structure of NaCl suddenly changes to the "cesium chloride" structure. This change in structure also causes the x-ray diffraction pattern to change, and we find that the first diffraction line is now at $\theta = 15.0°$. Identify this peak. (That is, find which planes in the "cesium chloride" structure reflect x rays at the smallest Bragg angle.)

Find the new lattice parameter a. Find the new density in g/cm^3. Answers: 4.84 Å, 3.42 g/cm^3, 2.98 Å, 3.67 g/cm^3.

At the beginning of this chapter, it was stated that we could use x-ray diffraction to determine the structure of crystals. We have examined in detail how to predict the position of the x-ray diffraction lines if the lattice is known. The relative intensity of the lines can be found by considering interference of waves reflected from points within a single primitive unit cell. The inverse problem is much more difficult. Given the positions of the diffraction lines, we want to be able to obtain the Bravais lattice of the crystal. Given the intensities of the lines, we want to be able to obtain the contents of the primitive unit cell, i.e., the basis of the lattice. The lattice plus its basis is the crystal structure. In practice, this is done by "guessing" what the crystal structure is. From this "guess," the position and intensity of the diffraction lines can be calculated and then compared with the measurements. If they do not agree, the "guess" is adjusted, and this procedure is repeated until satisfactory agreement is reached. As you might suspect, this procedure can be long and tedious. The general procedure of determining crystal structure from x-ray diffraction data has no simple solution.

CHAPTER 3

LATTICE VIBRATIONS

3-1 Harmonic Motion

In the previous two chapters, we have considered atoms to be fixed in position without motion. At room temperature, the atoms actually have a great deal of "thermal energy," and they vibrate quite vigorously about some average position.

To understand this motion, consider first of all the Na^+-Cl^- pair we discussed in Chapter 1. This pair of ions is at a position of equilibrium, $r = r_0$, when the net force is zero (see Fig. 3-1). If we pull this pair apart $(r > r_0)$, the force is then attractive and tries to bring the pair back to the equilibrium position. If we push this pair together $(r < r_0)$, the force is repulsive and again tries to bring the pair back to the

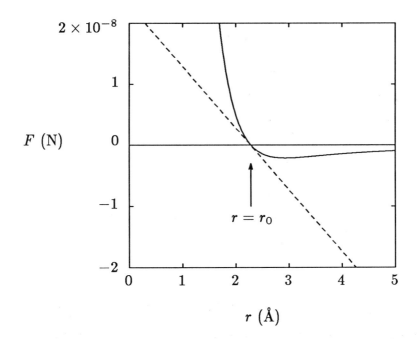

Fig. 3-1. The net force between a Na^+-Cl^- pair. Near $r = r_0$, this force is approximated by a straight line.

equilibrium position. The force between the Na^+-Cl^- pair is thus a **restoring force**. A displacement of the ions from their equilibrium position causes the force to try to restore them back to equilibrium.

If we pull the ions apart and let them go, they will oscillate about the equilibrium position. Consider what would happen if the restoring force F were simply proportional to the displacement, $r - r_0$, from the equilibrium position r_0, i.e.,

$$F = -\alpha(r - r_0). \tag{3-1}$$

We recognize this equation as **Hooke's Law**. The two ions behave as though they were connected by a spring of length r_0 and "spring constant" α. The resulting motion is **harmonic oscillation**,

$$r = r_0 + A \sin \omega t. \tag{3-2}$$

The amplitude A depends on how much energy the Na^+-Cl^- pair has, and the frequency ω depends on the magnitude of the spring constant α, as well as the masses of the two ions.

A restoring force such as that in Eq. (3-1) would be plotted as a straight line on a graph such as Fig. 3-1. As we can see, the actual restoring force for the Na^+-Cl^- pair is *not* a straight line and therefore does not obey Hooke's Law. However, *near* the equilibrium position ($r = r_0$), the force can be *approximated* by a straight line, as shown by the dashed line in Fig. 3-1. (That line passes through $F = 0$ at $r = r_0$ and has the same slope as F at $r = r_0$.) For *small* displacements from equilibrium, the force very nearly obeys Hooke's Law. Thus, small oscillations should be harmonic, as given in Eq. (3-2). We will use this approximation in the treatment that follows.

Problem 3-1. Find the spring constant α for a Na^+-Cl^- pair near the equilibrium position. Use the information given in Problem 1-21. Answer: 101 N/m.

Since we will consider only small oscillations, we can replace real forces between atoms by "springs," each with the appropriate spring constant α. As we saw above, a pair of atoms connected by a spring simply oscillates at some well defined frequency. In real crystals, there are forces in all three directions, and the motion of each atom is three-dimensional motion. However, great insight can be gained by studying a one-dimensional situation, and the analysis is much less complicated.

3-2 One-Dimensional Monatomic Lattice

Consider the one-dimensional lattice shown in Fig. 3-2. The atoms each have mass m and are connected to nearest neighbors by springs, each with spring constant α. When at rest in their positions of equilibrium, the atoms are separated by a distance a, the lattice parameter. Since all atoms are identical, we call this a **monatomic lattice**. We consider the lattice to be infinitely long, and we label the atoms with integers ($n - 1$, n, $n + 1$, $n + 2$ in Fig. 3-2). Let x_n be the position of the nth atom when all the atoms are at rest (at equilibrium). We can write

$$x_n = na. \tag{3-3}$$

Now consider the atoms in motion. Let u_n be the displacement of the nth atom from its position of equilibrium at x_n. The net force F_n on the nth atom is given by

$$F_n = -\alpha(2u_n - u_{n+1} - u_{n-1}). \tag{3-4}$$

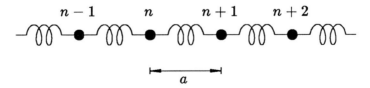

Fig. 3-2. One-dimensional monatomic lattice.

Problem 3-2. Show that the net force on the nth atom is given by Eq. (3-4) above.

Since the equilibrium positions x_n are time independent (by definition), we find that the acceleration of the nth atom is given by

$$a_n = \frac{d^2 u_n}{dt^2}. \qquad (3\text{-}5)$$

Using Newton's Law, $F = ma$, we obtain the equation of motion for the nth atom,

$$-\alpha(2u_n - u_{n+1} - u_{n-1}) = m\frac{d^2 u_n}{dt^2}. \qquad (3\text{-}6)$$

We can write down a similar equation for every other atom also. The solution to these equations is a traveling sine wave of the form

$$u_n = A\sin(kx_n - \omega t). \qquad (3\text{-}7)$$

If we put Eq. (3-7) into Eq. (3-6), we find that ω must be related to k in the following way:

$$\omega^2 = \omega_m^2 \sin^2(ka/2) \qquad (3\text{-}8)$$

or

$$\omega = \omega_m |\sin(ka/2)|, \qquad (3\text{-}9)$$

where

$$\omega_m = \sqrt{4\alpha/m}. \qquad (3\text{-}10)$$

When we take the square root of Eq. (3-8) to obtain ω, we want the positive root since frequency is defined to be a positive quantity. Thus, we have the absolute value of the sine function in Eq. (3-9). The relation between ω and k in Eq. (3-9) is called the **dispersion** of the lattice wave.

CHAPTER 3 LATTICE VIBRATIONS

Problem 3-3. Show that Eq. (3-7) is a solution to Eq. (3-6) if ω is given by Eq. (3-9). You will find Eq. (3-3) useful for working this problem.

In Fig. 3-3, we plot a lattice wave with a wavelength $\lambda = 8a$. The dashed lines represent x_n, the positions of the atoms when they are all at rest. The solid lines represent u_n, the displacements about x_n. We can see a wave moving to the right as the time t increases. (Notice that at $t = 0$, the atoms are "bunched up" around the 4th atom. At this instant, the 4th atom is at its equilibrium position. An instant later, when the 5th atom is at *its* equilibrium position, the atoms are bunched up around that atom. This bunching up of atoms is moving to the right. You may need to study the figure for a while to see this.) Notice that the movement of the atoms is parallel to the direction of the wave propagation. Such a wave is called **longitudinal**.

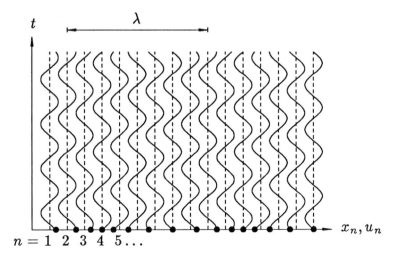

Fig. 3-3. One-dimensional lattice wave for $\lambda = 8a$ ($k = \pi/4a$). The dashed lines are x_n, the equilibrium positions of the atoms, and the solid lines are u_n, the displacements about x_n.

An alternate way to represent the lattice wave is shown in Fig. 3-4. The horizontal axis is x_n, the equilibrium positions of the atoms. The vertical axis is u_n, the displacements of the atoms from equilibrium. The values of u_n are shown for the wave "frozen" at some instant of time. This figure is *not* meant to imply that the motion of the atoms is perpendicular to the velocity of the wave. (Such a wave would be called **transverse**.) Rather, this is simply a graph showing displacement from equilibrium as a function of equilibrium position. The actual displacements u_n are *parallel* to the x axis. The sine curve in Fig. 3-4 moves to the right with increasing time t.

In Fig. 3-5, we plot the dispersion curve given by Eq. (3-9). Negative values of k describe waves moving in the $-x$ direction. From Eq. (2-6), we have

$$\omega = vk, \qquad (3\text{-}11)$$

where v is the velocity of the wave. In Fig. 3-5, this would be a straight line with slope v through the origin. The fact that the dispersion in Fig. 3-5 is *not* a straight line means that the velocity of a lattice wave depends on wavelength. The waves we are most familiar with (such as sound and light) have a velocity largely *independent* of wavelength. Waves in a lattice behave very differently. Also, we see that there exists a maximum frequency ω_m. There is a limit to the frequency at which the atoms can oscillate. *No* waves with a frequency greater than ω_m can exist in this lattice.

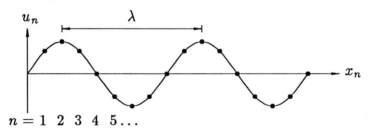

Fig. 3-4. Alternate graphical representation of the lattice wave in Fig. 3-3.

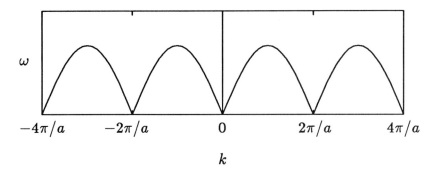

Fig. 3-5. The dispersion curve [Eq. (3-9)] for waves in a one-dimensional monatomic lattice.

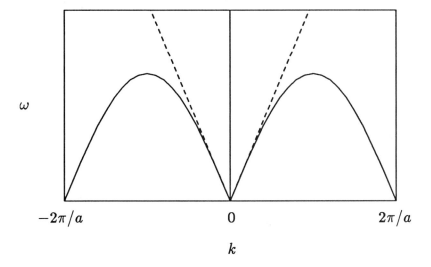

Fig. 3-6. Dispersion of a wave in a one-dimensional monatomic lattice. Dashed line shows the approximate dispersion near $k = 0$.

Let us examine the wave dispersion for very small values of k. Using the approximation that $\sin x \cong x$ for small x (x in units of radians), we find from Eq. (3-9) that

$$\omega \cong (\omega_m a/2)|k|. \tag{3-12}$$

Comparing with Eq. (3-11), we see that this wave has a velocity,

$$v = \omega_m a/2, \tag{3-13}$$

which is independent of wavelength. Eq. (3-12) is shown as a dashed line in Fig. 3-6. Note that the dashed line is a good approximation to the actual dispersion (the solid line) near $k = 0$.

Sound propagates through crystals as such a lattice wave. Since $\lambda = 2\pi/k$ by definition, very small values of k ($k \ll a^{-1}$) correspond to very long wavelengths ($\lambda \gg a$). A typical sound wave in a solid has a wavelength of the order of a meter or more. Thus, the velocity of sound in this crystal is given very accurately by Eq. (3-13) and is independent of wavelength.

At the other extreme, consider a wave with $k = \pi/a$ which has the maximum frequency possible ($\omega = \omega_m$). This wave has a wavelength $\lambda = 2a$. As seen in Figs. 3-7 and 3-8, half of the atoms move in one direction, and half move in the opposite direction.

3-3 First Brillouin Zone

The dispersion curve in Fig. 3-5 contains some redundancy. To see this, consider a wave with $k = -7\pi/4a$. We plot this wave in Fig. 3-9. If we compare Fig. 3-9 with Fig. 3-4, we see a striking similarity. Even though the wavelength is very different, the actual positions of the atoms are identical. In Fig. 3-10, we show the location of these two waves on the dispersion curve. We see that the wave numbers of these two waves differ by $\Delta k = 2\pi/a$ and that they have the same frequency ω. The motion of the atoms for these two waves is *identical*. The wave numbers $k = \pi/4a$ and $k = -7\pi/4a$ describe the same wave. (They are mathematically different,

CHAPTER 3 LATTICE VIBRATIONS 69

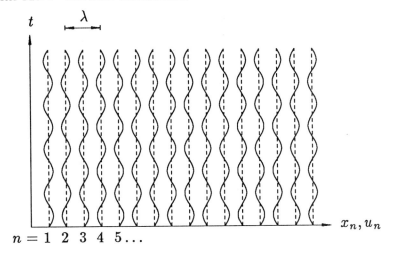

Fig. 3-7. One-dimensional lattice wave for $\lambda = 2a$ $(k = \pi/a)$.

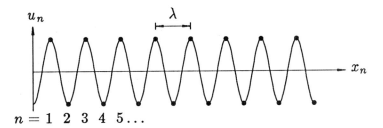

Fig. 3-8. Alternate graphical representation of the lattice wave in Fig. 3-7.

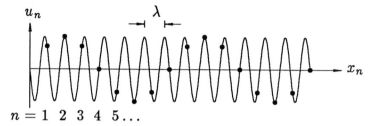

Fig. 3-9. One-dimensional lattice wave for $k = -7\pi/4a$.

but *physically* the same.) In general, any two wave numbers that differ by $2\pi/a$ describe the same physical wave in a one-dimensional lattice.

Problem 3-4. The wave in Fig. 3-9 is moving to the *left* ($k < 0$), whereas the wave in Fig. 3-4 is moving to the *right* ($k > 0$). Is the motion of the atoms really the same in these two waves? Explain.

Problem 3-5. Using Eqs. (3-7) and (3-9), show that any two waves whose wave numbers differ by $2\pi/a$ describe the same atomic motion.

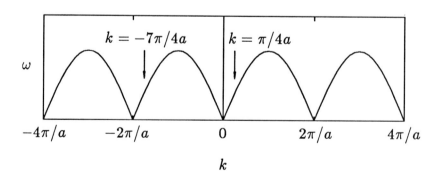

Fig. 3-10. Two waves separated by $\Delta k = 2\pi/a$.

Let us now consider the reciprocal one-dimensional lattice. The reciprocal lattice vectors **G** are defined by Eq. (2-16) to be solutions to the equation, $RG = 2\pi n$. (Since there is only one direction in one-dimensional space, $\mathbf{R} \cdot \mathbf{G} = RG$.) Each G must satisfy that equation simultaneously for *all* possible values of the direct lattice vectors **R**. Since R can be any multiple of a, then G must be a multiple of $2\pi/a$, that is, $G = n(2\pi/a)$. The k axis in Fig. 3-5 is the reciprocal space of a one-dimensional lattice. Points in reciprocal space which are separated by a reciprocal lattice vector **G** are *equivalent* points. Thus, we see that any two wave numbers which differ by $\Delta k = G = n(2\pi/a)$ are equivalent and must describe the same physical wave. (This is the same result that we obtained above.) All the information about possible lattice waves is thus contained within a unit cell in reciprocal space. The unit cell used for this purpose is the Wigner-Seitz cell which we introduced in Chapter 1 and is called the **first Brillouin zone**. For a one-dimensional lattice, the Wigner-Seitz cell of the reciprocal lattice is bounded by $k = \pm \pi/a$. This constitutes the first Brillouin zone of the one-

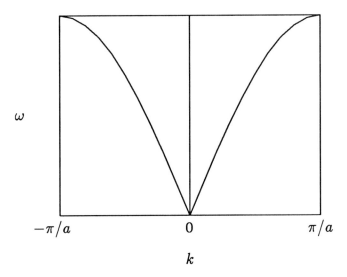

Fig. 3-11. Dispersion curve in the first Brillouin zone for the one-dimensional monatomic lattice.

dimensional lattice (see Fig. 3-11). Every wave in the lattice can be represented by some wave vector k in this zone.

3-4 One-Dimensional Diatomic Lattice

Next, let us see what happens in a one-dimensional lattice with *two* kinds of atoms (see Fig. 3-12). This lattice is called **diatomic**. Let the masses of the two kinds of atoms be m_1 and m_2. There are two atoms in each unit cell, and the distance between neighboring atoms is $\frac{1}{2}a$. The springs all have the same spring constant α as in the monatomic lattice. Also, as before, we label the atoms with integers $(2n-1, 2n, 2n+1, 2n+2, 2n+3$ in Fig. 3-12). However, here we label atoms of mass m_1 with *even* integers $(2n, 2n+2, $ etc.$)$ and atoms of mass m_2 with *odd* integers $(2n-1, 2n+1, 2n+3, $ etc.$)$. The equations of motion for the two kinds of atoms are given by

$$-\alpha(2u_{2n} - u_{2n+1} - u_{2n-1}) = m_1 \frac{d^2 u_{2n}}{dt^2} \qquad (3\text{-}14)$$

and

$$-\alpha(2u_{2n+1} - u_{2n+2} - u_{2n}) = m_2 \frac{d^2 u_{2n+1}}{dt^2}. \qquad (3\text{-}15)$$

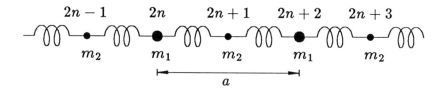

Fig. 3-12. One-dimensional diatomic lattice.

Problem 3-6. Show that Eqs. (3-14) and (3-15) are equations of motion for the atoms in the diatomic lattice.

The solution of these equations of motion is a traveling sine wave of the form,

$$u_{2n} = A_1 \sin(kx_{2n} - \omega t) \qquad (3\text{-}16)$$

and

$$u_{2n+1} = A_2 \sin(kx_{2n+1} - \omega t) \qquad (3\text{-}17)$$

All atoms of mass m_1 oscillate with amplitude A_1, and all atoms of mass m_2 oscillate with amplitude A_2. If we put these equations into Eqs. (3-14) and (3-15), we find

$$\omega^2 = \alpha\left(\frac{1}{m_1} + \frac{1}{m_2}\right) \pm \alpha\sqrt{\left(\frac{1}{m_1} + \frac{1}{m_2}\right)^2 - \frac{4\sin^2(ka/2)}{m_1 m_2}}. \qquad (3\text{-}18)$$

This dispersion curve is plotted in Fig. 3-13. We see that for every value of k, there are two values of ω, and thus there are two different lattice waves that exist for every wavelength. The two kinds of atoms oscillate with amplitudes which are related by

$$A_2 = A_1(1 - m_1\omega^2/2\alpha)\sec(ka/2). \qquad (3\text{-}19)$$

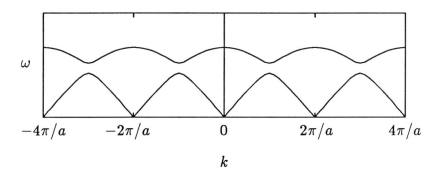

Fig. 3-13. Dispersion curves [Eq. (3-18)] for waves in a one-dimensional diatomic lattice.

Problem 3-7. Show that Eqs. (3-16) and (3-17) are solutions to Eqs. (3-14) and (3-15) if ω is given by Eq. (3-18). Show that A_1 and A_2 are related by Eq. (3-19).

As in the monatomic lattice, the dispersion curve is periodic in reciprocal space (the k axis). Any two wave vectors which differ by a reciprocal lattice vector, i.e., $\Delta k = G = n(2\pi/a)$, are equivalent and describe the same lattice wave. Thus, we only need to consider waves in the first Brillouin zone as shown in Fig. 3-14.

As an example, let us examine the waves for small k (large λ). From Eq. (3-18) we obtain two frequencies

$$\omega \cong ka\sqrt{\frac{\alpha}{2(m_1+m_2)}} \tag{3-20}$$

and

$$\omega \cong \sqrt{2\alpha\left(\frac{1}{m_1}+\frac{1}{m_2}\right)}. \tag{3-21}$$

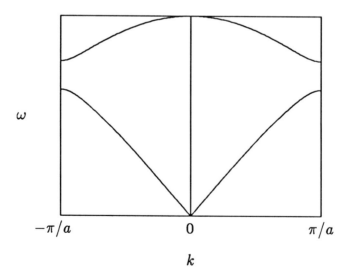

Fig. 3-14. Dispersion curve in the first Brillouin zone for the one-dimensional diatomic lattice.

CHAPTER 3 LATTICE VIBRATIONS   75

Problem 3-8. Show that for small k, the two values of ω from Eq. (3-18) are given approximately by Eqs. (3-20) and (3-21). You will need the relations, $\sin x \cong x$ and $\sqrt{1+x} \cong 1 + \frac{1}{2}x$, for small x.

Eq. (3-20) is the solution belonging to the lower curve in Fig. 3-14. The value of ω is very small on that curve near $k = 0$. From Eq. (3-19), we find that $A_2 \cong A_1$ for small k and small ω. Note that ω is proportional to k in Eq. (3-20). As in the case of the monatomic lattice, the dispersion curve of this wave near $k = 0$ is a straight line. Comparing with Eq. (3-11), we see that the velocity of this wave is given by

$$v = a\sqrt{\frac{\alpha}{2(m_1 + m_2)}} \qquad (3\text{-}22)$$

and is independent of wavelength. Sound propagates through the lattice with such a wave, and, for that reason, the lower curve in Fig. 3-14 is called the **acoustic branch**.

Eq. (3-21) is the solution belonging to the upper curve in Fig. 3-14. Notice that ω does *not* depend on k. This dispersion curve has zero slope at $k = 0$. Since these waves have a high frequency (large ω), they can interact with visible light (as we will discuss in Chapter 5). For this reason, the upper dispersion curve in Fig. 3-14 is called the **optical branch**.

Using $k = 0$ and the value of ω given by Eq. (3-21), we find from Eq. (3-19) that $A_2 = -(m_1/m_2)A_1$. The minus sign means that the atoms with mass m_2 are oscillating in an *opposite* direction as the atoms with mass m_1 (see Fig. 3-15). This lattice wave is very similar to the one shown in Fig. 3-7 for the monatomic lattice. Half of the atoms are oscillating one way, and half are oscillating the opposite way. However, in the diatomic lattice, the amplitude of oscillation of the two kinds of atoms are different (the magnitudes of A_1 and A_2 are different).

From Eq. (3-16), we see that when $k = 0$, the atomic displacement u_{2n} does not depend on x_{2n} and thus does not depend on n. At any given instant of time t, the displacements of all the atoms of mass m_1 are identical throughout the entire lattice. Similarly, from Eq. (3-17), we see that the same is true for the displacements of the atoms of mass m_2. The primitive unit cell of the diatomic lattice contains two atoms, one of mass m_1 and one of mass m_2. The position of the atoms within the unit cell at any given instant of time is thus identical for *every* unit cell in the lattice. This feature is evident in the lattice wave of Fig. 3-15. Since we see no change as we move from one unit cell to the next, the wavelength of the lattice wave is *infinite* ($k = 0$).

Problem 3-9. Compare Figs. 3-7 and 3-15. The wave in Fig. 3-7 has a wavelength equal to twice the distance between neighboring atoms. Would we be correct in saying that the wave in Fig. 3-15 also has a wavelength equal to twice the distance between neighboring atoms? What would be the wave number k of the lattice wave described that way?

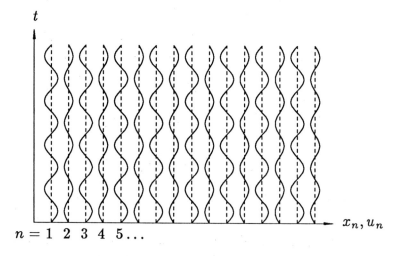

Fig. 3-15. Wave in one-dimensional diatomic lattice for $k = 0$ (large λ) in the optical branch. Here, we use $m_2 = \tfrac{2}{3} m_1$.

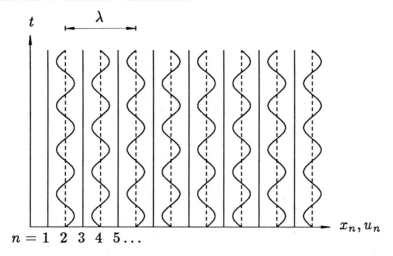

Fig. 3-16. Waves in the one-dimensional diatomic lattice for $\lambda = 2a$ ($k = \pi/a$) in the acoustic branch.

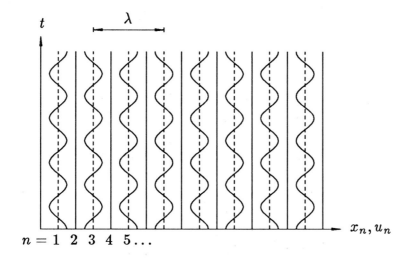

Fig. 3-17. Waves in the one-dimensional diatomic lattice for $\lambda = 2a$ ($k = \pi/a$) in the optical branch.

Next, let us examine the lattice waves at $k = \pi/a$ ($\lambda = 2a$). This is at the boundary of the first Brillouin zone. We will assume $m_1 > m_2$. From Eq. (3-18), we obtain

$$\omega = \sqrt{2\alpha/m_1} \quad \text{(acoustic branch)} \tag{3-23}$$

and

$$\omega = \sqrt{2\alpha/m_2} \quad \text{(optical branch)}. \tag{3-24}$$

For the acoustic branch, we have from Eq. (3-19) that $A_2 = 0$. The atoms of mass m_2 do *not* oscillate. Only the atoms of mass m_1 oscillate, as shown in Fig. 3-16. For the optical branch, we similarly have $A_1 = 0$. Only the atoms of mass m_2 oscillate, as shown in Fig. 3-17. We see that both waves (in the acoustic and optical branches) have the *same* wavelength but different frequencies. This arises from the difference in masses of the two kinds of atoms. The lighter atoms oscillate at a higher frequency than the heavy atoms.

Between the optical and acoustic branches, there is a range of frequencies where no lattice waves exist. Waves cannot propagate through the lattice at any of those frequencies. Thus, we

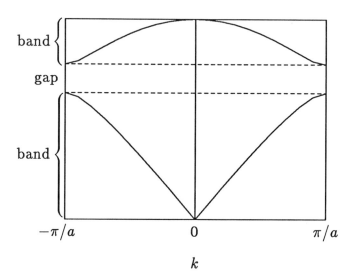

Fig. 3-18. Frequency bands separated by a gap.

CHAPTER 3 LATTICE VIBRATIONS

have two **bands** of allowed frequencies separated by a **gap** of forbidden frequencies, as shown in Fig. 3-18.

3-5 Three-Dimensional Crystals

In a three-dimensional crystal, we find many of the same features seen for the one-dimensional lattice. Traveling waves in three dimensions are written in the form,

$$\mathbf{u}_n = \mathbf{A}\sin(\mathbf{k}\cdot\mathbf{r}_n - \omega t), \tag{3-25}$$

where $\mathbf{r}_n$ is the equilibrium position of the nth atom, and $\mathbf{u}_n$ is the displacement from equilibrium. The wave travels in the direction of the wave vector $\mathbf{k}$. As in the one-dimensional case, each lattice wave in three dimensions can be described by some wave vector $\mathbf{k}$ in the first Brillouin zone, which is the Wigner-Seitz cell of the reciprocal lattice.

For example, consider copper (Cu). The direct lattice is fcc. The reciprocal lattice is bcc. The first Brillouin zone is the Wigner-Seitz cell of the bcc lattice shown in Fig. 3-19. The frequency ω is a function of position in the three-dimensional reciprocal space. We usually plot the dispersion curve along some particular *line* in reciprocal space. In Fig. 3-20, we show the dispersion curves for Cu along three different directions from the origin out to the first Brillouin zone boundary. These three lines are shown as dashed lines in Fig. 3-19. We see that the dispersion curves for the three directions are different. The behavior of a lattice wave depends on the direction of propagation in the crystal.

Problem 3-10. Find the length of the line from the center of the first Brillouin zone out to the boundary along the [100] and [111] directions in Cu. You will find Fig. 1-16 helpful. Answer: 1.74 Å^{-1}, 1.51 Å^{-1}.

Also, in three-dimensional waves, there are three different polarizations possible for a given direction of propagation $\mathbf{k}$.

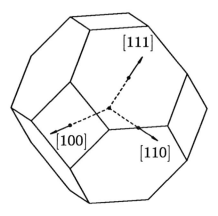

Fig. 3-19. The first Brillouin zone of the fcc lattice.

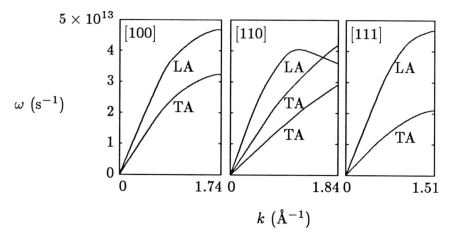

Fig. 3-20. Dispersion curves in copper (Cu) along the dashed lines shown for the first Brillouin zone in Fig. 3-19. Data are from E. C. Svensson, B. N. Brockhouse, and J. M. Rowe, *Phys. Rev.* **155**, 619 (1967).

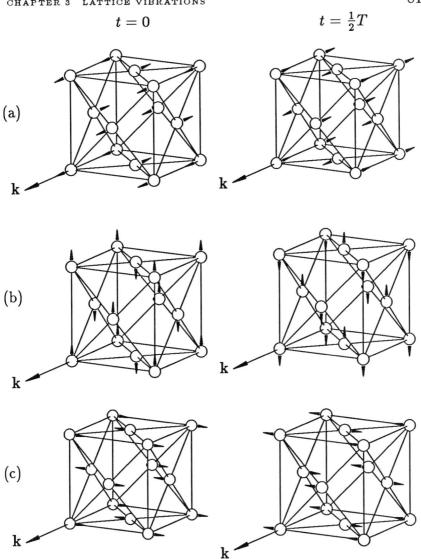

Fig. 3-21. Displacement of atoms in copper (Cu) for a lattice wave in the [100] direction with $k = 1.74$ Å^{-1} (the boundary of the first Brillouin zone). The displacements are frozen at two different times, $t = 0$ and $t = \frac{1}{2}T$ (one-half period of the oscillation). The three possible polarizations are shown: (a) longitudinal, (b) transverse, and (c) transverse.

In Fig. 3-21 are shown lattice waves in Cu propagating in the [100] direction (the direction of **k**). The atomic displacements are shown for two instants of time, $t = 0$ and $t = \frac{1}{2}T$ (one-half period of the oscillation). The longitudinal wave (**A** parallel to **k**) is shown in (a), the two possible transverse waves (**A** perpendicular to **k**) are shown in (b) and (c). In Cu, all the lattice waves are acoustic (the lattice is monatomic), and thus in Fig. 3-20 we label the curves LA for longitudinal acoustic and TA for transverse acoustic. In the [100] and [111] directions, both TA branches in Cu have the same dispersion curve and are thus drawn on top of each other as one curve. The two transverse waves in Fig. 3-21 have the same frequency, whereas the longitudinal wave has a higher frequency.

Using the dispersion curves, we can find the velocity of sound waves from the slope of the curve near $k = 0$ as in Fig. 3-6. We see that in general the velocity of a sound wave depends on its polarization as well as its direction in the crystal.

Problem 3-11. From the dispersion curve in Fig. 3-20, find the velocity of the LA sound wave in copper (Cu) along the [100] direction. Answer: 4600 m/s.

Consider next a crystal of potassium bromide (KBr). The direct lattice (Bravais) is fcc, and the reciprocal lattice is bcc, as in Cu. The dispersion curves are shown in Fig. 3-22 for three directions in the crystal. The crystal is diatomic, and thus we observe optical branches, labeled LO and TO for the longitudinal and transverse polarizations, respectively. Acoustic branches go to zero at $k = 0$. Optical branches do not. As in the one-dimensional diatomic lattice, we observe a frequency gap between the acoustic and optical branches.

As a last example, consider the dispersion curves for diamond (C) in Fig. 3-23. The direct lattice (Bravais) is fcc, and the reciprocal lattice is bcc, as in the previous two examples of Cu and KBr. We see optical branches in Fig. 3-23 even

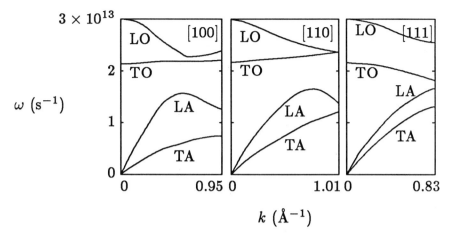

Fig. 3-22. Dispersion curves in potassium bromide (KBr) in the first Brillouin zone. Data are from A. D. B. Woods, B. N. Brockhouse, and R. A. Cowley, *Phys. Rev.* **131**, 1025 (1963).

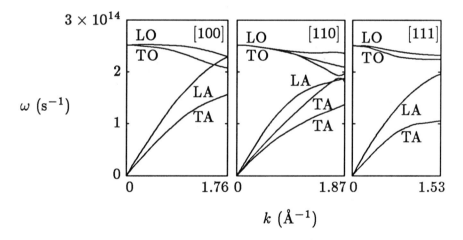

Fig. 3-23. Dispersion curves for diamond (C) in the first Brillouin zone. Data are from J. L. Warren, G. Dolling, and R. A. Cowley, *Phys. Rev.* **158**, 805 (1967).

though the crystal contains only one kind of atom. Remember that in the diamond structure, there are two atoms in the primitive unit cell. Not all atoms in the crystal are at equivalent positions. There are *two* kinds of atoms in the crystal, even though they are all C atoms. Thus, a diamond crystal is actually "diatomic" as far as the lattice waves are concerned.

Let us return to the question posed at the beginning of this chapter. What kind of atomic motion is present in a crystal? Lattice waves are continually being excited by thermal energy in the crystal. Atoms are oscillating about their positions of equilibrium. At room temperature in most crystals, lattice waves with frequencies throughout all the bands of allowed frequency are present and are superimposed on each other. A given atom is oscillating at many different frequencies simultaneously.

How does this atomic motion affect x-ray diffraction? The "average" position of an atom is its equilibrium position associated with a Bravais lattice point. We still observe Bragg reflections of x rays. Bragg's Law for x-ray diffraction still holds true, on the average. However, the lattice vibrations cause the *intensity* of the reflected x rays to be diminished.

3-6 Thermal Expansion

As a final topic in this chapter, let us consider thermal expansion. As we increase the temperature of solid, it expands. The lattice parameter a becomes larger. This phenomenon is caused by increased atomic motion due to a greater amount of thermal energy. To see how this causes expansion of the lattice, consider the Na^+-Cl^- pair which we discussed at the beginning of this chapter. For oscillations of small amplitude, we found that the restoring force follows Hooke's Law and the Na^+-Cl^- pair undergoes harmonic oscillation like two masses connected by a spring. Note, however, that two masses connected by a spring always oscillate *symmetrically* about their equilibrium position r_0. The average distance between the two ions is equal to r_0, no matter how great the amplitude of the oscillation (as long as the oscillation is still harmonic). In a crystal, the lattice parameter a depends on the *average* distance between atoms.

If the oscillations of atoms in a crystal were truly harmonic, then the average distance between atoms would not change with temperature, and there would be *no* thermal expansion.

To be able to explain the phenomenon of thermal expansion, we must consider the actual interatomic forces. They do *not* follow Hooke's Law. The actual oscillations of atoms are **anharmonic** (not harmonic). We see in Fig. 3-1 that the restoring force between the Na^+-Cl^- pair is *weaker* for $r > r_0$ than for $r < r_0$. Thus, the oscillation is *not* symmetric about the point of equilibrium r_0. The amplitude of the oscillation for $r > r_0$ is larger than that for $r < r_0$, and the average distance r between the two ions is consequently *greater* than r_0.

To treat this phenomenon quantitatively, we consider the potential energy U given by

$$U = -\int_{r_0}^{r} F\, dr, \qquad (3\text{-}26)$$

where we have chosen the lower limit of the integral to be r_0 so that U would be zero at $r = r_0$. If F obeys Hooke's Law, as in Eq. (3-1), we obtain the familiar expression,

$$U = \tfrac{1}{2}\alpha(r - r_0)^2, \qquad (3\text{-}27)$$

for the potential energy of a spring.

In Fig. 3-24, we plot the actual potential energy (the solid line) near the point of equilibrium as well as the harmonic approximation (the dashed line) given by Eq. (3-27). The total energy E of the Na^+-Cl^- pair is the sum of the kinetic and potential energies and must be *conserved*. Thus, at the two extremes of the motion where the kinetic energy is zero, the potential energy U must be equal to the total energy E. To find the two extremes of the motion, then, we simply set $U = E$ and find the two solutions of this equation.

For example, if the total energy were $E = 1.0 \times 10^{-19}$ J, we see from Fig. 3-24 that $U = E$ at two points, $r \cong 1.9$ Å and 2.9 Å. The two ions oscillate between these two values of r. The average value of r during the oscillation is given *approximately* by the midpoint between the two extremes. In this

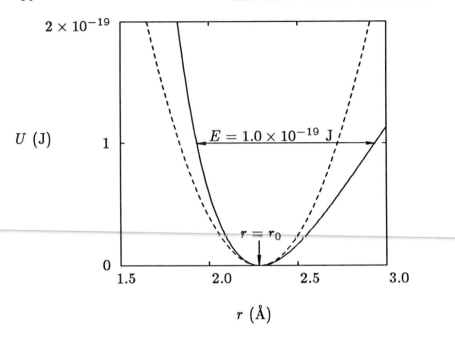

Fig. 3-24. Potential energy of a Na$^+$-Cl$^-$ pair. The dashed line is the harmonic approximation. The horizontal line shows the oscillation for a total energy equal to 1.0×10^{-19} J.

case, then, the average value of r is about 2.4 Å, which is about 0.1 Å greater than r_0. The motion causes the average distance between the Na$^+$ and Cl$^-$ ions to increase by about 4%.

It is evident from Fig. 3-24 that if the Na$^+$-Cl$^-$ pair had even a greater amount of energy, the average value of r would be greater. This is the physical basis of thermal expansion of a crystal. At greater temperatures, the atoms have more thermal energy and oscillate with a greater amplitude. Because the oscillations are anharmonic, the average distance between the atoms increases, and the crystal expands.

Problem 3-12. At room temperature, a typical Na$^+$-Cl$^-$ pair has a thermal energy equal to about 1.0×10^{-20} J. Using the harmonic approximation, find the amplitude of oscillation for a pair of ions with this energy. Use the results of Problem 3-1. Answer: 0.14 Å.

Problem 3-13. Consider a Na^+-Cl^- pair. Find the potential energy as a function of r. (Use the information given in Problem 1-21.) If the ions have a thermal energy equal to 1.0×10^{-20} J, find the approximate value of the average distance between two ions. (Do not use the harmonic approximation here.) By what percentage is this value greater than r_0? Answer: 2.30 Å, about 1%.

CHAPTER 4

CLASSICAL MODEL OF METALS

4-1 Conduction Electrons

The behavior of electrons in solids is by far the most important topic in solid state physics and will be the subject of discussion for the remainder of this book. We start our discussion with the study of metals since many of the features of the behavior of electrons in metals can be described with classical physics and can thus be easily visualized. In this chapter, then, we will begin our study of electrons in solids by examining the classical model of metals.

A metallic solid contains a large number of electrons which are free to move throughout the solid. These electrons give rise to the most distinguishing property of a metal: electrical conductivity. These electrons are thus called **conduction electrons**.

As an example, consider a very simple metal, sodium (Na). Each Na atom has 11 electrons. One of the electrons is loosely bound to the atom and is thus easily removed. This electron is called a **valence electron**. The other 10 electrons are much more tightly bound to the atom and are called **core** electrons. When Na atoms form a crystal, all the valence electrons are stripped from the atoms and become conduction electrons. The Na atoms, each with one missing electron, become positively charged Na^+ ions.

Since each Na atom contributes one conduction electron to the metal, the number of conduction electrons must be equal to the number of atoms. Thus, the density of conduction electrons is equal to the density of Na atoms. Since the crystal structure of Na is bcc, there are two Na atoms in each conventional unit cell of volume a^3. The density n of conduction electrons in Na is thus

$$n = \frac{2}{a^3} = \frac{2}{(4.30 \times 10^{-10} \text{ m})^3} = 2.52 \times 10^{28}/\text{m}^3.$$

Some atoms have more than one valence electron. The number of valence electrons in an atom is called the **valence** Z of the atom. In a metal, all valence electrons become conduction electrons. Thus, the density n of the conduction electrons in a metal is simply Z times the density of the atoms. In Appendix 4 are listed a number of metals and their valence.

Problem 4-1. Find the density of conduction electrons in aluminum (Al). Answer: 1.82×10^{29} /m^3.

4-2 Electric Current

Conduction electrons move throughout the metal. They continually interact with the atoms as well as with each other. We will use a simple model to take into account the most important features of these interactions. We treat the conduction electrons as point masses which are moving about in a "box" filled with "obstacles" of some kind. The electrons frequently "collide" with these obstacles. Following each collision, an electron acquires a new velocity, both in direction and magnitude. In between collisions, the electron is free of any interaction and moves in a straight line at constant velocity. The path of an electron in this model may look something like that shown in Fig. 4-1. This model is not very realistic, but it allows us to make a number of very simple calculations, as will be shown below. In the absence of any external force, the velocities **v** of the conduction electrons are randomly distributed in direction. The average velocity is zero since for every electron with velocity **v**, there is an electron with velocity −**v**.

Let us now apply an electric field $\mathcal{E}$ to the metal. The force **F** on each conduction electron is given by

$$\mathbf{F} = -e\mathcal{E}, \tag{4-1}$$

where e is the magnitude of the charge of an electron. From Newton's Law, $\mathbf{F} = m\mathbf{a}$, we find the acceleration of the electron to be

$$\mathbf{a} = -(e/m)\mathcal{E}. \tag{4-2}$$

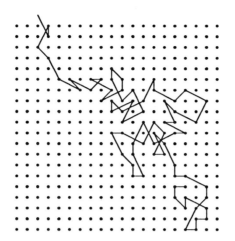

Fig. 4-1. Path of a conduction electron moving through a metal. The dots represent obstacles of some kind.

In our model, the electron is accelerated in between collisions, but at each collision, the electron loses all memory of this acceleration so that it leaves the collision with some random velocity again. The extra kinetic energy which the electron acquires during its acceleration is given up to the obstacle. The displacement **r** traveled by an electron between two consecutive collisions is given by

$$\mathbf{r} = \tfrac{1}{2}\mathbf{a}t^2 + \mathbf{v}_0 t, \tag{4-3}$$

where t is the time between the two collisions and $\mathbf{v}_0$ is the "initial velocity," i.e., the velocity of the electron immediately following the first of the two collisions. The average displacement $\langle \mathbf{r} \rangle$ traveled by an electron between any two consecutive collisions is given by

$$\langle \mathbf{r} \rangle = \tfrac{1}{2}\mathbf{a}\langle t^2 \rangle. \tag{4-4}$$

The average of the last term in Eq. (4-3) is zero since we assumed that the velocity $\mathbf{v}_0$ immediately following a collision is random in direction. Also note that **a** is constant and does not need to be averaged. Let us define τ to be the *average time between collisions* of a given electron, i.e., $\tau \equiv \langle t \rangle$. The average velocity of the electron is simply given by $\langle \mathbf{v} \rangle = \langle \mathbf{r} \rangle / \tau$. This velocity $\langle \mathbf{v} \rangle$ is called the **drift velocity** $\mathbf{v}_d$ of the electron. From Eq. (4-4), we obtain

$$\mathbf{v}_d = \tfrac{1}{2}\mathbf{a}\langle t^2 \rangle / \tau. \tag{4-5}$$

We must now evaluate $\langle t^2 \rangle$. One might naively think that $\langle t^2 \rangle = \tau^2$. However, this is not true. The average of the square of some quantity is generally *greater* than the square of its average. Thus, we expect that $\langle t^2 \rangle$ is greater than τ^2.

Problem 4-2. Consider the numbers 1,2,4,8,16. Find the square of their average and the average of their squares. Answer: 38.44, 68.2.

The exact relationship between $\langle t^2 \rangle$ and τ^2 depends on the statistical distribution of t. A discussion of this topic is beyond the scope of this book, so I will just tell you the answer. From classical physics, we find that in this case, $\langle t^2 \rangle = 2\tau^2$. Putting this into Eq. (4-5) and using the expression for $\mathbf{a}$ in Eq. (4-2), we finally obtain

$$\mathbf{v}_d = -(e\tau/m)\mathbf{\mathcal{E}}. \tag{4-6}$$

Note here that under the influence of an electric field $\mathcal{E}$, the electrons acquire a *non-zero average* velocity in a direction opposite to $\mathcal{E}$. The electrons are moving from one end of the metal to the other. This constitutes an electric **current** I.

4-3 Conductivity

Consider a metallic wire of length L and cross-sectional area A (see Fig. 4-2). We apply a voltage V across the length of the wire. The resulting electric field $\mathcal{E} = V/L$ in the wire causes a current I to flow. In a time interval Δt, a conduction electron moves an average distance $v_d \Delta t$. Thus, all the conduction electrons contained within a volume $Av_d \Delta t$ pass by a given

CHAPTER 4 CLASSICAL MODEL OF METALS

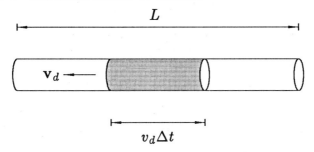

Fig. 4-2. Wire of length L and cross-sectional area A carrying an electrical current.

point on the wire during that time interval. For an electron density n, we see that $nAv_d\Delta t$ electrons pass by that point during that time interval. Since the charge of an electron is $-e$, a total charge $\Delta q = -enAv_d\Delta t$ passes by. The current $I \equiv \Delta q/\Delta t$ is thus equal to $-enAv_d$. The **current density** J is defined to be the current per unit cross-sectional area, i.e., $J = I/A$. So we finally have

$$\mathbf{J} = -ne\mathbf{v}_d. \qquad (4\text{-}7)$$

The current density **J** is a vector quantity and is in the opposite direction of $\mathbf{v}_d$ since the charge of an electron is negative.

Problem 4-3. Consider a wire made of copper (Cu). If the wire is 2 mm in diameter and carries 10 A of current, find the drift velocity of the conduction electrons. Answer: 0.234 mm/s.

Combining Eqs. (4-6) and (4-7), we obtain

$$\mathbf{J} = (ne^2\tau/m)\mathcal{E} \qquad (4\text{-}8)$$

or

$$\mathbf{J} = \sigma\mathcal{E}, \qquad (4\text{-}9)$$

where
$$\sigma = ne^2\tau/m. \tag{4-10}$$

The quantity σ is called the electrical **conductivity** of the metal. It does not depend on A or L, the dimensions of the wire. The conductivity is a property of the metal itself. Its value is different for each type of metal. The conductivity of a number of metals is given in Appendix 4. (Some tables give the **resistivity** ρ which is simply equal to $1/\sigma$.) Using experimentally measured values of σ, we can use Eq. (4-10) to calculate the average time τ between collisions of the conduction electrons.

Problem 4-4. Find the average time between collisions of a conduction electron in sodium (Na). Answer: 3.35×10^{-14} s.

Since σ is independent of $\mathscr{E}$, we see from Eq. (4-9) that the current density is proportional to the applied electric field. This is known as **Ohm's Law**. A more familiar form of Ohm's Law is
$$V = IR, \tag{4-11}$$
where R is the **resistance** of the wire. Using $E = V/L$ and $J = I/A$, we obtain from Eqs. (4-9) and (4-11),
$$R = L/A\sigma. \tag{4-12}$$

Note that resistance *does* depend on the dimensions of the wire.

Problem 4-5. Find the resistance of a copper wire 2 mm in diameter and 10 m long. Repeat for an aluminum wire of the same dimensions. Answer: 0.0532 Ω, 0.0845 Ω.

The expression for σ in Eq. (4-10) implies that metals with a higher concentration of conduction electrons will have a higher conductivity. This is not generally true. For example, aluminum (Al) has nearly three times the concentration of conduction electrons as copper (Cu) and yet has a *smaller* conductivity. Eq. (4-10) would force us to conclude that the average time τ between collisions of electrons in Al is more than three times shorter than in Cu. But there is no apparent reason why τ should be so much shorter in Al than in Cu. (In fact, it is not.) This difficulty arises from deficiencies in the classical model of metals and will be discussed again in a later chapter.

4-4 Hall Effect

Next, let us place a metallic solid in a magnetic field **B**. If we drive an electrical current through the metal, the field **B** exerts an *average* force **F** on the moving electrons,

$$\mathbf{F} = -e\mathbf{v}_d \times \mathbf{B}. \qquad (4\text{-}13)$$

In Fig. 4-3, we show a piece of metal with current flowing to the right. The electrons are moving to the left (since they have negative charge) with an average velocity $\mathbf{v}_d$. The field **B** points out of the page. Thus, the force **F** from Eq. (4-13) is downward, as shown. This force causes electrons to move downward, and the bottom part of the metal becomes negatively charged. A deficiency of electrons in the upper part of the metal causes it to become positively charged (Na^+ ions are left behind). The positive charge on the upper part of the metal and the negative charge on the lower part together create an electric field called

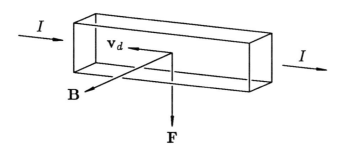

Fig. 4-3. Current flowing through a piece of metal in a magnetic field.

the **Hall field** $\mathscr{E}_H$ which points downward and thus exerts an *upward* force on the electrons, partially canceling the force of the magnetic field **B**. The net vertical force on each electron is given by

$$\mathbf{F} = -e(\mathbf{v}_d \times \mathbf{B}) - e\mathscr{E}_H. \quad (4\text{-}14)$$

Electrons continue to be forced downward by **B** until $\mathscr{E}_H$ becomes large enough that its force completely cancels the force of **B**. At that point, the system is in equilibrium (at least in the vertical direction), and the net vertical force on each electron is zero. Setting $\mathbf{F} = 0$ in Eq. (4-14), we obtain

$$\mathscr{E}_H = -\mathbf{v}_d \times \mathbf{B}. \quad (4\text{-}15)$$

Using Eq. (4-7), we can write this as

$$\mathscr{E}_H = \frac{1}{ne}\mathbf{J} \times \mathbf{B} \quad (4\text{-}16)$$

or

$$\mathscr{E}_H = -R_H \mathbf{J} \times \mathbf{B}, \quad (4\text{-}17)$$

where

$$R_H = -1/ne. \quad (4\text{-}18)$$

The electric field $\mathscr{E}_H$ causes a small but detectable voltage across the metal. This phenomenon is called the **Hall effect**. The quantity R_H is called the **Hall coefficient**.

Problem 4-6. Using Eq. (4-18), calculate the Hall coefficient for sodium (Na). Repeat for copper (Cu). Answer: -2.48×10^{-10} m^3/C, -0.734×10^{-10} m^3/C.

The experimental values of R_H for a number of metals are given in Appendix 4. We find there that, for example, $R_H = -2.50 \times 10^{-10}$ m^3/C for Na and -0.55×10^{-10} m^3/C for Cu. As we can see from the problem above, the theory works fine for Na but not so well for Cu. Things get even worse for some other metals. The Hall coefficients for iron (Fe) and zinc (Zn), for example, are *positive*. They have the wrong sign. The classical model of metals cannot explain this.

Problem 4-7. Consider a slab of copper 0.100 mm thick, 1.00 mm wide, and 10.0 mm long. (a) If we drive a current of 1.00 A down the length of the slab, what is the current density J? (b) If we then put the slab in a magnetic field $B = 1.00$ T with the field perpendicular to the 1 mm × 10 mm face, what Hall field E_H will be produced? Use the actual Hall coefficient in Appendix 4 instead of the classical value, $R_H = -1/ne$. (c) What Hall voltage will we observe across the slab? Answer: 1.00×10^7 A/m^2, 5.5×10^{-4} V/m, 0.55 μV.

4-5 Cyclotron Resonance

Even when no current is being driven through a metallic solid, the conduction electrons are continually in motion. The instantaneous force of a magnetic field **B** on these electrons is

$$\mathbf{F} = -e\mathbf{v} \times \mathbf{B}, \tag{4-19}$$

where **v** is now the instantaneous velocity. This force is in a direction perpendicular to **v** and thus causes the electrons to go in circles (see Fig. 4-4). An object moving in a circle of radius r has an acceleration v^2/r. Using $F = ma = mv^2/r$ and assuming that **v** is perpendicular to **B**, we obtain from Eq. (4-19),

$$evB = mv^2/r. \tag{4-20}$$

From this equation, we can solve for the angular velocity $\omega = v/r$ of the electron and obtain

$$\omega = eB/m. \tag{4-21}$$

Note that ω is independent of v. All the conduction electrons, regardless of their velocity, move in circles with the *same* angular velocity. If we shine electromagnetic radiation on the

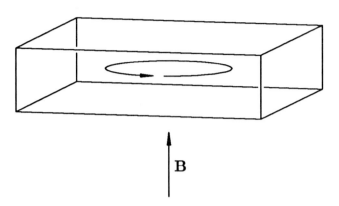

Fig. 4-4. A magnetic field causes conduction electrons to move in circles.

metal, we find that the radiation is strongly absorbed by the metal when the frequency of the radiation is equal to ω. This phenomenon is called **cyclotron resonance**.

Problem 4-8. Consider a field $B = 1.00$ T. Using Eq. (4-21), find the frequency (in Hz) and wavelength of radiation at the cyclotron resonance in a metal placed in that field. Answer: 2.80×10^{10} Hz, 1.07 cm.

As can be seen in the problem above, cyclotron resonance occurs for microwave radiation. In practice, cyclotron resonance can only be observed if the electron can do a few circles before colliding, i.e., $\omega\tau > 1$. This condition can only be met for very pure samples at very low temperatures. If we actually measure experimentally the cyclotron resonance in various metals, we find that ω does not follow Eq. (4-21) exactly. For some metals, the experimental value of ω is smaller. For other metals, it is larger. This is again an effect which the classical model of metals cannot adequately explain.

CHAPTER 5

WAVES AND PARTICLES

5-1 Photons

In 1905, certain experimental evidence led A. Einstein to postulate that the energy of an electromagnetic wave is *quantized*. By this, we mean that the energy can only take on certain *discrete* values. He postulated that these allowed values of energy are $E = nh\nu$, where ν is the frequency of the wave, h is a proportionality constant called **Planck's constant**, and n is any positive integer. A wave can only have an energy which is a multiple of $h\nu$. It cannot have any energy which is in between these values. The value $h\nu$ is called a **quantum** of energy. Energy is not continuous but is *quantized*. The energy quantum $h\nu$ of electromagnetic radiation is called a **photon**. A wave of total energy $nh\nu$ contains n photons.

Electromagnetic waves also carry momentum as well as energy. A wave with total energy E carries a momentum $p = E/c$, where c is the speed of light. Thus, if energy is quantized, then momentum must also be quantized. Each photon carries a momentum equal to $p = h\nu/c = h/\lambda$. An electromagnetic wave behaves in many respects like a collection of "particles" (photons), each with energy E and momentum p given by

$$E = h\nu \tag{5-1}$$

and

$$p = h/\lambda. \tag{5-2}$$

In terms of the angular frequency ω and wave vector **k** of the wave, we find that

$$E = \hbar\omega \tag{5-3}$$

and

$$\mathbf{p} = \hbar\mathbf{k}, \tag{5-4}$$

where $\hbar \equiv h/2\pi$ ($\hbar$ is pronounced "h bar").

Problem 5-1. Consider visible light of wavelength 500 nm. Find the energy and momentum of the photons in this light wave. Answer: 2.48 eV, 1.32×10^{-27} kg·m/s.

The actual existence of photons was first firmly established experimentally by the **Compton effect**, which was discovered in 1923. If we shine electromagnetic radiation (such as x rays) on some solid material, we find scattered x rays emerging from the material with a wavelength which is *longer* than that of the incident x rays. This effect can be understood to be the result of *inelastic* collisions between photons of the x rays and electrons in the material. An incident photon gives up some of its energy and momentum to an electron and emerges from the collision with less energy and momentum and thus a longer wavelength.

The wavelength of the emerging photon can be calculated using conservation of energy and momentum. Let us consider the electron to be initially at rest. Let E and $\mathbf{p}$ be the final kinetic energy and momentum, respectively, of the electron after

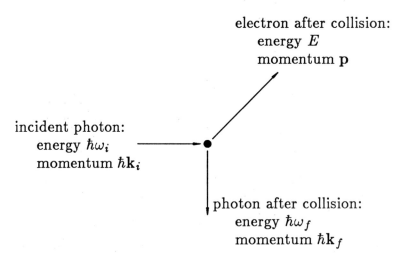

Fig. 5-1. The Compton effect. Collision between a photon and an electron.

CHAPTER 5   WAVES AND PARTICLES   101

the collision. Then, from conservation of energy and momentum, we have
$$\hbar\omega_i = \hbar\omega_f + E \tag{5-5}$$
and
$$\hbar\mathbf{k}_i = \hbar\mathbf{k}_f + \mathbf{p}, \tag{5-6}$$
where the subscripts, i and f, refer to the initial and final states of the photon, respectively (see Fig. 5-1). Careful experimental measurements of the Compton effect were found to be in complete agreement with these equations.

Problem 5-2. Consider a collision between a photon and an electron. The electron is initially at rest. After the collision, the photon emerges in a direction 90° from its original direction (as shown in Fig. 5-1). The initial wavelength of the photon is 1.542 Å (an x ray), and its final wavelength is 1.566 Å. (a) Find the energy of the photon before and after the collision. (b) From conservation of energy, find the velocity of the electron after the collision. (c) Find the momentum of the photon before and after the collision. (d) Using the velocity of the electron found from conservation of energy, show that the total momentum is also conserved in this collision. Answer: 8.05 keV, 7.92 keV, 6.8×10^6 m/s, 4.29×10^{-24} kg·m/s, 4.22×10^{-24} kg·m/s.

Problem 5-3. Consider a collision between a photon and an electron as described in Problem 5-2 (same initial conditions) except that the photon emerges in a direction 180° from its original direction (it goes back where it came from). (a) Using conservation of energy and momentum, find the final wavelength of the photon. (b) Find the final velocity of the electron. Answer: 1.590 Å, 9.29×10^6 m/s.

5-2 Phonons

The energy of other types of waves has also been found to be quantized. A case of great importance in solid state physics is that of lattice waves, which we discussed in Chapter 3. The quantum of energy in a lattice wave is given by $E = \hbar\omega$ as in Eq. (5-3) for an electromagnetic wave. This energy quantum of a lattice wave is called a **phonon**. A lattice wave of total energy $n\hbar\omega$ contains n phonons. Just like the photons, we can consider phonons to be "particles." Other particles (such as electrons, neutrons, and even photons) which enter a crystal can "collide" with these phonons, exchanging energy and momentum.

Actually, lattice waves do not carry any momentum themselves. However, whenever another particle "collides" with a phonon, it behaves as though the phonon did have momentum $\mathbf{p} = \hbar\mathbf{k}$, as in Eq. (5-4) for photons. The quantity $\hbar\mathbf{k}$ is called the **crystal momentum** of the phonon.

5-3 Inelastic Scattering of Neutrons

As an example, consider a neutron which enters a crystal and interacts with a lattice wave of frequency ω and wave vector $\mathbf{k}$. The neutron may either *take* energy from the lattice wave, or it may *give* energy to the lattice wave. In either case, the neutron emerges from the crystal with a different energy and momentum than it had when it entered the crystal. The amount of energy exchanged between the neutron and the lattice wave must be a multiple of $\hbar\omega$ since the energy of the lattice wave is quantized. We will consider here the case where only *one* quantum $\hbar\omega$ of energy is exchanged.

If the neutron *takes* this energy $\hbar\omega$ from the lattice wave, then we say that the neutron **absorbs** one phonon. From conservation of energy, we see that the energy E of the neutron increases by $\hbar\omega$. There also exists a "crystal momentum" conservation law for this collision. The momentum of the neutron must increase by $\hbar\mathbf{k}$, the crystal momentum of the absorbed phonon. Thus, for phonon absorption,

$$E_f = E_i + \hbar\omega \tag{5-7}$$

and

$$\mathbf{p}_f = \mathbf{p}_i + \hbar\mathbf{k}, \tag{5-8}$$

CHAPTER 5   WAVES AND PARTICLES   103

where i and f refer to the initial and final states of the neutron.

If the neutron *gives* energy to the lattice wave, we say that the neutron **emits** one phonon. The neutron must therefore *lose* energy $\hbar\omega$ and momentum $\hbar\mathbf{k}$, and we have for phonon emission,

$$E_f = E_i - \hbar\omega \tag{5-9}$$

and

$$\mathbf{p}_f = \mathbf{p}_i - \hbar\mathbf{k}. \tag{5-10}$$

Like Bragg's Law for x-ray diffraction, these equations are quite restrictive. These scattered neutrons emerge from the crystal only in certain directions.

As an example, consider inelastic scattering of neutrons from copper (Cu). In Fig. 5-2 is shown a typical energy spectrum of scattered neutrons from a sample of copper. These data were taken with the neutron spectrometer set up as shown in Fig. 5-3. The Cu sample is at the origin. The x, y, z axes refer to the [100], [010], and [001] directions, respectively, in the Cu crystal. The crystal is oriented so that the incident neutrons are in the x-y plane at an angle 14.0° with the y axis, as shown. The neutron detector is also placed in the x-y plane, but at an angle 45.5° with the y axis, as shown. The detector measures the energy of the scattered neutrons along that direction. The spectrum in Fig. 5-2 shows the distribution of energy among the neutrons scattered in that direction.

The incident neutrons have an energy $E_i = 5.6 \times 10^{-21}$ J. We see a prominent peak in Fig. 5-2 at this energy. The scattered neutrons with this energy have been *elastically* scattered by the sample (the energy of the neutrons did not change) and are of no interest to us here. The other two peaks in Fig. 5-2 arise from neutrons which have been *inelastically* scattered by phonons. Since the energy is higher, they *absorbed* a phonon. Let us consider in detail the peak marked "A" in Fig. 5-2.

Problem 5-4. Using conservation of energy, find the frequency ω of the phonons absorbed by the neutrons in peak "A" of Fig. 5-2. Answer: 4.3×10^{13} s^{-1}.

104 CHAPTER 5 WAVES AND PARTICLES

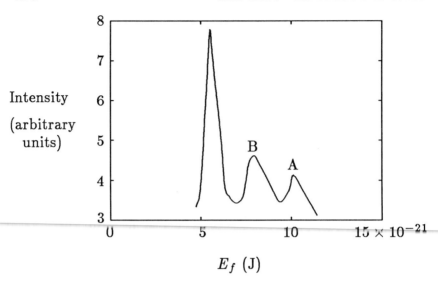

Fig. 5-2. Inelastic scattering of neutrons from copper. Reprinted with permission from *J. Phys. Chem. Solids* **23**, J. Sosnowski and J. Kuzubowski, "Phonon dispersion relations for copper single crystal in the [100] direction," Copyright 1962, Pergamon Press, Ltd.

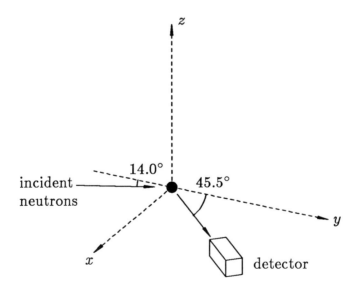

Fig. 5-3. Directions of incident and scattered neutrons for the spectrum in Fig. 5.2.

Problem 5-5. Using the conservation of momentum, find the magnitude and direction of the wave vector **k** of the phonons absorbed by the neutrons in peak "A" of Fig. 5-2. Answer: 4.9 Å^{-1}, approximately along the [100] direction.

From the results of the above problems, we find that the neutrons in peak "A" of Fig. 5-2 absorbed energy from a lattice wave propagating along the [100] direction with $\omega = 4.3 \times 10^{13}$ s^{-1} and $k = 4.9$ Å^{-1}. In Problem 3-10, we found that the first Brillouin zone in Cu only extended to $k_{max} = 1.74$ Å^{-1} along the [100] direction. Our phonon here has a wave vector *outside* the first Brillouin zone. As we saw in Chapter 3, any lattice wave can be described by a wave vector **k** *inside* the first Brillouin zone. This is due to the fact that all wave vectors **k** which differ by a reciprocal lattice vector **G** describe the *same* lattice wave. Thus, by adding an appropriately chosen reciprocal lattice vector **G** to the wave vector of our phonon we can find its wave vector in the first Brillouin zone.

Problem 5-6. Consider the phonon in Problem 5-5. Find its wave vector **k** (magnitude and direction) in the first Brillouin zone. Answer: 1.4 Å^{-1} in the [100] direction.

Note that the phonon can be described by $k = 4.9$ Å^{-1} or $k = 1.4$ Å^{-1} along the [100] direction. It is the *same* phonon. The wave vector **k** of a phonon is not uniquely given but may have any of the values **k** + **G**. Consequently, the crystal momentum of a phonon is also not uniquely given but may have any of the values $\hbar\mathbf{k} + \hbar\mathbf{G}$. This means that the momentum which a neutron gains when it absorbs a phonon (or loses when it emits a phonon) may have any of the values $\hbar\mathbf{k} + \hbar\mathbf{G}$.

From the information provided in peak "A" of Fig. 5-2, we have found that there must be a lattice wave propagating along the [100] direction in Cu with wave number $k = 1.4$ Å^{-1} and frequency $\omega = 4.3 \times 10^{13}$ s^{-1}. This is a point on a dispersion curve. From Fig. 3-20, we see that this lattice wave is a point on the LA branch of the dispersion curve for the [100] direction. By varying the orientation of the crystal and the position of the detector, we could experimentally obtain other points on these dispersion curves as well. Dispersion curves for lattice waves in crystals are usually experimentally obtained in this manner, point by point, from neutron scattering data.

Problem 5-7. Consider the phonons absorbed by the neutrons in peak "B" in Fig. 5-2. Find their frequency ω and wave vector **k** (magnitude and direction) in the first Brillouin zone. From Fig. 3-20, would you guess that these phonons are longitudinal or transverse? Answer: 2.3×10^{13} s^{-1}, 1.1 Å^{-1}, 28° from the [100] direction.

5-4 Inelastic Scattering of Photons

Light waves can also be inelastically scattered by lattice waves in a crystal. As in neutron scattering, photons in the light wave may either absorb or emit a phonon. Similar to Eqs. (5-7) through (5-10), we have for phonon absorption,

$$\omega_f = \omega_i + \omega \tag{5-11}$$

and

$$\mathbf{k}_f = \mathbf{k}_i + \mathbf{k}, \tag{5-12}$$

and for phonon emission,

$$\omega_f = \omega_i - \omega \tag{5-13}$$

and

$$\mathbf{k}_f = \mathbf{k}_i - \mathbf{k}, \tag{5-14}$$

where ω_i and $\mathbf{k}_i$ refer to the incident photon, ω_f and $\mathbf{k}_f$ refer to the scattered photon, and ω and $\mathbf{k}$ refer to the phonon.

A photon in visible light has much more energy than any phonon in the crystal. Thus, $\omega \ll \omega_i$, and the absorption or emission of a phonon changes the frequency of the photon only by a relatively very small amount, i.e., $\omega_f \cong \omega_i$. Since the wave number of the light wave in the crystal is proportional to its frequency, we consequently also have $k_f \cong k_i$.

Typically, in a light-scattering spectrometer, the detector is placed so that it receives light which is scattered at an angle 90° from the incident beam (see Fig. 5-4). We shine a beam of light through the crystal and view it from the side. From the triangles in Fig. 5-4, we immediately obtain

$$k = k_i\sqrt{2}, \qquad (5\text{-}15)$$

since $k_f \cong k_i$ and the angle between $\mathbf{k}_i$ and $\mathbf{k}_f$ is 90°.

A photon in visible light has a wave number much smaller than the typical dimensions of the first Brillouin zone. This means that the wave vector $\mathbf{k}$ of the absorbed or emitted phonon must lie near the center ($\mathbf{k} = 0$) of the first Brillouin zone. Near $\mathbf{k} = 0$, there are two kinds of phonons: optical phonons with a rather large frequency ω independent of $\mathbf{k}$ and acoustic phonons with a very small frequency ω proportional to k. Photons scattered by optical phonons will have a frequency ω_f well-separated from the frequency ω_i of the incident photons, whereas photons scattered from acoustic phonons will have a frequency ω_f much closer to ω_i. When the scattering involves optical phonons, it is called **Raman scattering**. When it involves acoustic phonons, it is called **Brillouin scattering**.

Let us consider Raman scattering first. Since the frequency ω of the optical phonons near $\mathbf{k} = 0$ is independent of $\mathbf{k}$, the shift in frequency, $\Delta\omega = \omega_f - \omega_i$, of the scattered photons will be independent of the frequency ω_i of the incident photons and also independent of the crystal orientation. The shift $\Delta\omega$ will be equal to $+\omega$ for phonon absorption and $-\omega$ for phonon emission. The Raman spectrum for diamond is shown in Fig. 5-5. A large central unshifted peak ($\Delta\omega = 0$) is observed for photons which are elastically scattered. Two other peaks can also be seen, one on each side of the central peak. One of these peaks is due to phonon absorption ($\omega_f = \omega_i + \omega$), and the other is due to phonon emission ($\omega_f = \omega_i - \omega$). The frequency shifts of these two peaks can be accurately measured and are found to be $\Delta\omega = \pm 2.513 \times 10^{14}$ s^{-1}. This is the frequency ω of the lattice wave in the optical branch near $k = 0$, in agreement with the dispersion curves for diamond in Fig. 3-23.

Problem 5-8. Find the wavelengths of the inelastically scattered photons in Fig. 5-5. Answer: 5836 Å, 6910 Å.

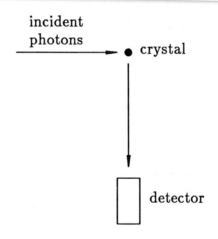

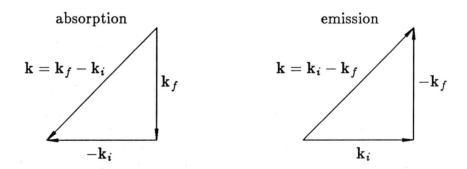

Fig. 5-4. Detection of inelastically scattered photons from a crystal. The two triangles show the conservation of crystal momentum for phonon absorption and emission.

CHAPTER 5   WAVES AND PARTICLES

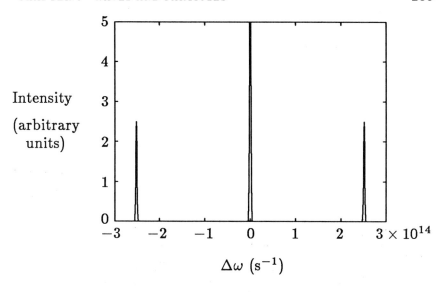

Fig. 5-5. Raman spectrum of diamond using laser light with $\lambda = 6328$ Å ($\omega_i = 2.979 \times 10^{15}$ s^{-1}). See, for example, S. A. Solin and A. K. Ramdas, *Phys. Rev. B* **1**, 1687 (1970).

Next let us consider Brillouin scattering. Since the frequency ω of the acoustic phonons near $\mathbf{k} = 0$ is proportional to k along any given direction, the shift in frequency, $\Delta\omega = \omega_f - \omega_i$, of the scattered photons in this case will depend both on the frequency ω_i of the incident photons and on the crystal orientation. The Brillouin spectrum for diamond is shown in Fig. 5-6. We see here two peaks on each side of the central unshifted peak since there are two kinds of acoustic phonons, longitudinal and transverse. (In the Raman spectrum, there was only one peak on each side since the optical longitudinal and transverse phonons in diamond have the same frequency near $\mathbf{k} = 0$.) In Brillouin scattering, the frequency shifts due to phonon emission are called Stokes shifts, and those due to phonon absorption are called anti-Stokes shifts.

The laser light used in Fig. 5-6 has a wavelength $\lambda = 4880$ Å in air. In the crystal, the speed of light is *slower* than c and thus its wavelength is *shorter* (the frequency remains the same). To obtain the wavelength of a light wave in a crystal, we simply divide its wavelength in air by the **index of refraction** of the crystal. The index of refraction of diamond is 2.4173. Thus, the wavelength of the laser light in the diamond crystal is 2019 Å. We must use this wavelength when considering conservation of crystal momentum during the collision.

Problem 5-9. Find the wave vector **k** (magnitude and direction) of the phonons in the Brillouin spectrum of Fig. 5-6. The incident photons are in the [110] direction, and the scattered photons are in the [$\bar{1}$10] direction. ($\bar{1}$ means -1.) Answer: 4.40×10^{-3} Å^{-1} along the [0$\bar{1}$0] direction.

Problem 5-10. Using the frequency shifts $\Delta\omega$ in Fig. 5-6 and using the result of Problem 5-9, find the velocity of sound in

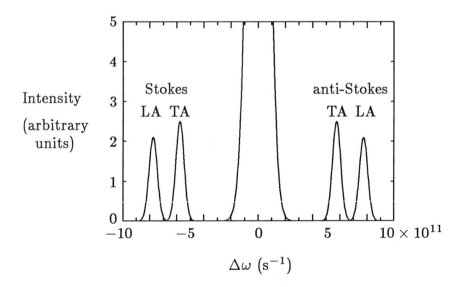

Fig. 5-6. Brillouin spectrum of diamond using laser light with $\lambda = 4880$ Å ($\omega_i = 3.863 \times 10^{15}$ s^{-1}). The incident photons are in the [110] direction, and the scattered photons are in the [$\bar{1}$10] direction. Data are from M. H. Grimsditch and A. K. Ramdas, *Phys. Rev. B* **11**, 3139 (1975).

diamond along the [100] direction. Answer: 18000 m/s for longitudinal waves and 13000 m/s for transverse waves.

Problem 5-11. Find the difference in wavelength of the photons in the central peak and those in one of the TA peaks of Fig. 5-6. Answer: 0.7 Å.

5-5 Wave-like Properties of Particles

We have seen that waves sometimes behave like particles with momentum $p = \hbar k = h/\lambda$. In 1924, L. de Broglie postulated that particles may sometimes behave like *waves*. He said that the "wavelength" λ of a particle should be related to its momentum p by

$$\lambda = h/p. \tag{5-16}$$

This wavelength is called the **de Broglie wavelength** of the particle.

Problem 5-12. Find the de Broglie wavelength of an electron which has a kinetic energy equal to 100 eV. Answer: 1.23 Å.

As can be seen from the result of the above problem, electrons with moderate energy have a de Broglie wavelength equal to that of x rays. If de Broglie's postulate is correct, electrons should exhibit wave behavior like we usually observe for x rays. For example, electrons should be "diffracted" by crystals. In 1927, C. J. Davisson and L. H. Germer in the United States and G. P. Thomson in Scotland experimentally confirmed de Broglie's postulate by observing electron diffraction in crystals. They found that the diffraction followed Bragg's Law, using the de Broglie wavelength for λ.

Problem 5-13. Consider diffraction of electrons from the (100) planes in aluminum (Al). Find the minimum energy (in units of eV) of the electrons that would permit us to observe reflection at a Bragg angle of 10°. Answer: 306 eV.

Since the first electron-diffraction experiments, scientists have found a great deal of use for particle diffraction by crystals. A very important application is the determination of crystal structure by **neutron diffraction**. Neutron diffraction has some distinct advantages over x-ray diffraction. Neutrons are mainly scattered by atomic nuclei, whereas x rays are mainly scattered by electrons. As a result, "light" atoms (such as H, C, N, and O) which have very few electrons scatter neutrons much more strongly than x rays. Also, a neutron has a magnetic dipole and is therefore very sensitive to magnetic effects in a crystal. Magnetic fields have no effect on x rays at all. Neutron diffraction sometimes reveals details of a crystal structure which cannot be obtained from x-ray diffraction. On the other hand, neutrons are more difficult to produce than x rays, and thus neutron diffraction is a more expensive technique.

A neutron diffraction pattern for powdered iron (Fe) is shown in Fig. 5-7. These data were taken by a method very similar to the x-ray diffraction technique shown in Fig. 2-15. Instead of photographic film to record the diffraction lines, a neutron detector is placed at some angle 2θ, as shown in Fig. 5-8. (As in x-ray diffraction, θ is the Bragg angle.) The intensity of the diffracted neutrons must be obtained at each angle by placing the detector at that angle. The diffraction pattern in Fig. 5-7 was actually obtained by measuring the neutron intensity at about 80 different angles, each corresponding to a different position of the detector.

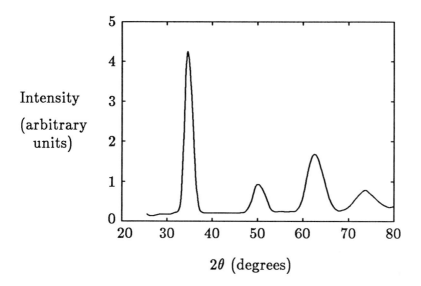

Fig. 5-7. Neutron diffraction pattern for powdered iron. Data are from C. C. Shull, E. O. Wollan, and W. C. Koehler, *Phys. Rev.* **84**, 912 (1951).

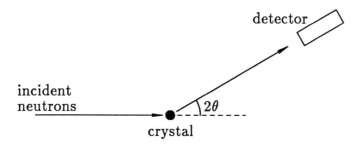

Fig. 5-8. Neutron diffraction spectrometer.

Problem 5-14. The crystal structure of iron (Fe) is bcc. Which planes in the crystal diffract the neutrons at the smallest angle? (See Problem 2-19.) From the position of the first peak in Fig. 5-7, find the de Broglie wavelength and the energy of the neutrons. Identify the other three peaks in Fig. 5-7 as well (i.e., find the crystal planes which diffracted the neutrons). Answer: 1.19 Å, 9×10^{-21} J.

CHAPTER 6

QUANTUM MECHANICS

6-1 Wave Functions

In the last chapter, we saw how particles sometimes behave like waves. Newton's law is inadequate for describing this wave-like nature of particles. In this chapter, we will introduce the formal mathematical theory, called **quantum mechanics**, which takes the place of Newton's law in describing the behavior of small particles.

First, we must answer a fundamental question: what is oscillating? In all other types of waves we have discussed, we could identify *something* that oscillates. In a lattice wave, the atoms oscillate. In an electromagnetic wave, electric and magnetic fields oscillate. A moving particle behaves like a wave with the de Broglie wavelength. We call this wave a **matter wave**. What is oscillating in this wave? It is *not* the particle itself that is oscillating. In fact, none of the particle's physical properties are oscillating either. It is some abstract, non-physical property of the particle which is oscillating. We call this abstract property the **wave function**, denoted by ψ. The wave function of a particle is generally a function of position (x, y, z) and time and contains the complete information about the particle (its position, momentum, energy, etc.). We cannot directly observe the wave function itself. It is purely an abstract quantity. But we *can* measure physical quantities (such as the particle's position, momentum, energy, etc.) which depend on the wave function and thus indirectly obtain information about it.

The wave function of a particle is generally a *complex* function, so we need to first introduce complex numbers. Square roots of negative numbers are called **imaginary numbers**. We denote the square root of -1 with the symbol i, that is, $i \equiv \sqrt{-1}$. Numbers which do not contain any imaginary part are called **real numbers**. All *physical* quantities are measured with real numbers. When real numbers are combined with imaginary numbers, we get **complex numbers** which can always be written in the form, $z = x + iy$, where x and y are real numbers. The **absolute value** of a complex number z is denoted by $|z|$ and is defined to be

$$|z| = \sqrt{x^2 + y^2}. \tag{6-1}$$

We often see imaginary numbers used as an exponent of the number e. Consider the exponential function $\exp(ix) \equiv e^{ix}$, where x is a real number. We can write the series expansion,

$$\exp(ix) = 1 + (ix) + \tfrac{1}{2}(ix)^2 + \tfrac{1}{6}(ix)^3 + \cdots$$
$$= \sum_{n=0}^{\infty} \frac{1}{n!}(ix)^n. \tag{6-2}$$

Noting that $i^2 = -1$ and $i^3 = -i$, we can write this equation as,

$$\exp(ix) = (1 - \tfrac{1}{2}x^2 + \cdots) + i(x - \tfrac{1}{6}x^3 + \cdots)$$
$$= \sum_{n=0}^{\infty} \frac{1}{n!}(-1)^n x^{2n} + i \sum_{n=0}^{\infty} \frac{1}{n!}(-1)^n x^{2n+1} \tag{6-3}$$
$$= \cos x + i \sin x.$$

Using Eq. (6-1), we find that the absolute value of $\exp(ix)$ is given by

$$|\exp(ix)| = 1. \tag{6-4}$$

Problem 6-1. Derive Eq. (6-4) from Eqs. (6-1) and (6-3).

With this introduction to complex numbers, we now write down the wave function of a free particle in one-dimensional space (we only consider one coordinate x),

$$\psi = A \exp(ikx - i\omega t). \tag{6-5}$$

Using Eq. (6-3), we see that this can also be written as

$$\psi = A\cos(kx - \omega t) + iA\sin(kx - \omega t). \qquad (6\text{-}6)$$

This is simply a traveling wave with wave number k and frequency ω. It has both a real and imaginary part, which give this traveling wave some special properties, as we shall see. The quantities, k and ω, are related to the particle's momentum p and energy E by the relations, $p = \hbar k$ and $E = \hbar\omega$, which we introduced in the last chapter.

We can obtain p and E from the wave function in Eq. (6-5) by taking appropriate derivatives:

$$-i\hbar\frac{\partial\psi}{\partial x} = p\psi, \qquad (6\text{-}7)$$

$$i\hbar\frac{\partial\psi}{\partial t} = E\psi. \qquad (6\text{-}8)$$

The symbols, $\partial\psi/\partial x$ and $\partial\psi/\partial t$, are special kinds of derivatives called **partial derivatives**. In a partial derivative, we take the derivative with respect to *one* of the variables, treating the other variables as constants. Thus, $\partial\psi/\partial x$ is a derivative of ψ with respect to x, treating t as a constant, and $\partial\psi/\partial t$ is a derivative of ψ with respect to t, treating x as a constant.

Problem 6-2. Show that Eqs. (6-7) and (6-8) are correct for the wave function in Eq. (6-5).

6-2 Schroedinger's Equation

For a free particle (no forces acting on it), the energy is equal to the kinetic energy, $E = \tfrac{1}{2}mv^2 = p^2/2m$, and from Eqs. (6-7) and (6-8), we have

$$-\frac{\hbar^2}{2m}\frac{\partial^2\psi}{\partial x^2} = i\hbar\frac{\partial\psi}{\partial t}. \qquad (6\text{-}9)$$

If the particle is *not* free but is subject to some force, the function in Eq. (6-5) is no longer the wave function of the particle.

However, the correct wave function still obeys an equation like Eq. (6-9). We only need to include the effect of the force on the particle. First, we identify $E = \hbar\omega$ as the *total* energy of the particle. Then we introduce the potential energy U which includes all the effects of the force on the particle. From conservation of energy, the total energy must equal the sum of the kinetic energy and potential energy. Thus we need to add the potential energy to the left-hand side of Eq. (6-9), and we obtain

$$-\frac{\hbar^2}{2m}\frac{\partial^2 \psi}{\partial x^2} + U\psi = i\hbar\frac{\partial \psi}{\partial t}. \qquad (6\text{-}10)$$

This is called **Schroedinger's equation**. Given the potential energy, we can, in principle, solve Schroedinger's equation for ψ. Since ψ contains all the information about the position and motion of the particle, we can consider Schroedinger's equation as the equation of motion in quantum mechanics which replaces Newton's law, $F = ma$, in classical physics. For a free particle, we have $U = 0$, and the solution to Schroedinger's equation is simply Eq. (6-5).

Note that we really have not *derived* Schroedinger's equation. We have simply used some plausibility arguments to show how it at least makes sense. Actually, Schroedinger's equation *cannot* be "derived," any more than Newton's law, $F = ma$, can be derived. Its validity rests upon experimental evidence. Schroedinger's equation has proven to give an accurate description of the behavior of particles, as far as we have been able to experimentally determine.

6-3 Wave Function of a Free Particle

Let us now make a connection between ψ and the physical properties of a particle. The probability that a particle is at some position x is proportional to $|\psi|^2$. The function $|\psi|^2$ is called the **probability function** of the particle's position. For example, we see that for a free particle whose wave function is given by Eq. (6-5), we have, using Eq. (6-4),

$$|\psi|^2 = A^2, \qquad (6\text{-}11)$$

independent of position and time. (A is just the amplitude of the wave and is a constant.) It is thus equally probable for this particle to be *anywhere*. This is not a very realistic description of a particle since we usually have *some* idea where the particle is.

We can construct a more realistic wave function of a free particle by adding together traveling waves of the form of Eq. (6-5) with different wave numbers and frequencies. Any wave function which is a sum of functions which satisfy Schroedinger's equation will also satisfy Schroedinger's equation.

Problem 6-3. Let ψ_1 and ψ_2 be two different solutions of Schroedinger's equation. Show that $\psi = \psi_1 + \psi_2$ is also a solution of Schroedinger's equation.

Consider the following wave function:

$$\psi = A \int_{k_1}^{k_2} \exp(ikx - i\omega t)\,dk. \tag{6-12}$$

This is simply the sum of an infinite number of wave functions like Eq. (6-5) with wave numbers k between k_1 and k_2. We can show that ψ is a solution to Schroedinger's equation for a free particle.

Problem 6-4. Show that Eq. (6-12) is a solution to Schroedinger's equation for a free particle.

The frequency ω is a function of k, and the integral in Eq. (6-12) cannot be done. However, we can make the following approximation. If the interval from k_1 to k_2 is small, we can expand ω about an average value of the wave number:

$$\omega(k) \cong \omega(\bar{k}) + \left.\frac{d\omega}{dk}\right|_{k=\bar{k}} (k - \bar{k}), \tag{6-13}$$

where $\bar{k}$ is the average value of k over the interval and is given by

$$\bar{k} = \tfrac{1}{2}(k_1 + k_2). \tag{6-14}$$

If we define

$$\bar{\omega} = \omega(\bar{k}) \tag{6-15}$$

and

$$v_g = \left.\frac{d\omega}{dk}\right|_{k=\bar{k}}, \tag{6-16}$$

then

$$\omega = \bar{\omega} + (k - \bar{k})v_g. \tag{6-17}$$

Putting this expression for ω into Eq. (6-12), we can now do the integral, and we obtain

$$\psi = 2A\frac{\sin[(x - v_g t)\Delta k/2]}{x - v_g t}\exp(i\bar{k}x - i\bar{\omega}t), \tag{6-18}$$

where

$$\Delta k = k_2 - k_1. \tag{6-19}$$

Problem 6-5. Using the approximation for ω in Eq. (6-17), do the integral in Eq. (6-12) and obtain the result, Eq. (6-18).

The probability function $|\psi|^2$ of the particle's position is given by

$$|\psi|^2 = 4A^2\frac{\sin^2[(x - v_g t)\Delta k/2]}{(x - v_g t)^2} \tag{6-20}$$

and is plotted in Fig. 6-1 for $t = 0$. The maximum of $|\psi|^2$ (the central peak in Fig. 6-1) is at $x - v_g t = 0$ or $x = v_g t$. We can thus identify v_g as the velocity of the central peak in $|\psi|^2$. The whole function $|\psi|^2$ moves along the x direction with velocity v_g, as shown in Fig. 6-2. The particle will most likely be found in a location where $|\psi|^2$ is large, which is near the central peak shown in Fig. 6-1. Since this central peak is

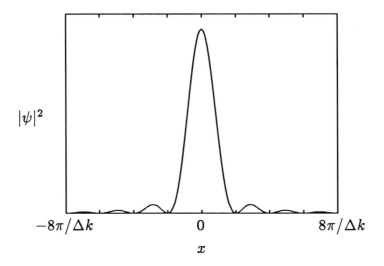

Fig. 6-1. The probability function in Eq. (6-20) at $t = 0$.

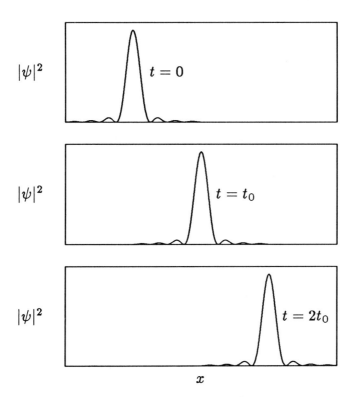

Fig. 6-2. The probability function at times $t = 0$, t_0, and $2t_0$.

traveling with velocity v_g, the particle is also. In other words, the wave function ψ in Eq. (6-12) represents a particle moving with velocity v_g. This velocity v_g is called the **group velocity** of the wave.

For a free particle, we have

$$\omega = \frac{\hbar}{2m}k^2. \qquad (6\text{-}21)$$

The group velocity of the particle is thus

$$v_g = \frac{d\omega}{dk} = v, \qquad (6\text{-}22)$$

which is the actual velocity of the particle. The velocity of the wave itself is called the **phase velocity** v_{ph}. It is given by the usual relation for a wave, $v_{ph} = \lambda\nu = \omega/k$. We find that the phase velocity of a matter wave is equal to $\frac{1}{2}v$. The phase velocity is only half the actual velocity of the particle.

Problem 6-6. Using $p = \hbar k$ and $E = \hbar\omega$, show that $\omega = (\hbar/2m)k^2$ for a free particle.

Problem 6-7. Using Eqs. (6-21) and (6-22), show that the group velocity of the wave function for a free particle is the actual velocity of the particle.

Problem 6-8. Show that the phase velocity of a matter wave is equal to half the velocity of the particle.

The meaning of the group and phase velocities is best explained by illustration. The real part of ψ (denoted Re ψ) in Eq. (6-18) is given by

$$\text{Re } \psi = 2A\frac{\sin\left[(x - v_g t)\Delta k/2\right]}{x - v_g t}\cos(\bar{k}x - \bar{\omega}t). \qquad (6\text{-}23)$$

We plot Re ψ in Fig. 6-3 for various times t. As we can see, the traveling wave, $\cos(\bar{k}x - \bar{\omega}t)$, is contained within an "envelope"

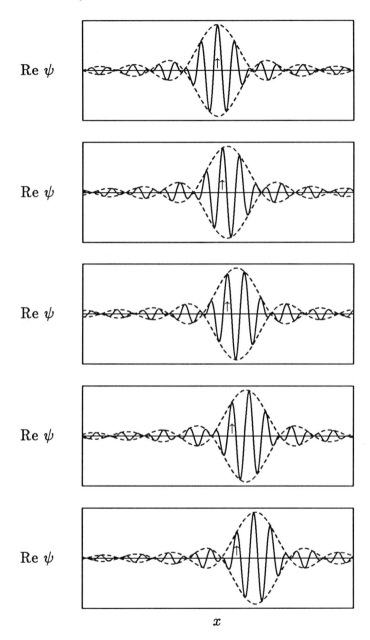

Fig. 6-3. The real part of the wave function at various times t. The envelope given by the dashed line is moving with group velocity v_g. The vertical arrow is moving with phase velocity v_{ph}.

(the dashed lines). The wave is traveling at velocity $v_{ph} = \frac{1}{2}v$. The envelope is traveling at velocity $v_g = v$, which is twice as fast.

By combining wave functions such as Eq. (6-5) which extend over all space, we were able to construct a wave function such as Eq. (6-18) which is restricted to an "envelope." In so doing, we gave up information about the momentum. We combined wave functions for particles which have momentum anywhere between $\hbar k_1$ and $\hbar k_2$. Thus, the uncertainty Δp in the momentum is approximately given by $\Delta p \cong \hbar \Delta k$. The location of the particle is also somewhat uncertain. We only know that it is most likely within the central peak of the envelope. So the uncertainty Δx in the position of the particle is approximately equal to the width of the central maximum or $\Delta x \cong 1/\Delta k$ (to within an order of magnitude). The product of Δx and Δp is

$$\Delta x \Delta p \cong \hbar, \qquad (6\text{-}24)$$

which is independent of x and p. The above equation is called the **Heisenberg uncertainty principle**. The position and momentum of a particle cannot be known simultaneously to any degree of accuracy desired.

In our example, in order to fix the location of the particle more accurately (reduce Δx), we must make the central peak of $|\psi|^2$ in Fig. 6-1 narrower, which we can only do by *increasing* Δk and thus increasing the uncertainty in the particle's momentum.

In the wave function given by Eq. (6-5), there is no uncertainty in k, and thus $\Delta p = 0$. However, as we saw in Eq. (6-11), it is equally probable for the particle to be anywhere, and thus $\Delta x = \infty$.

Problem 6-9. Consider an electron with kinetic energy equal to 100 eV. If we know this value is accurate to within 1 eV, how accurately can we know the location of the electron? Answer: approximately to within 40 Å.

6-4 Particle in a Box

As an example of the application of quantum mechanics to a real problem, let us consider a particle trapped in a "box" of length L. (In the next chapter, we will use this model for electrons in a metal.) Inside the box, there are no forces acting on the particle. It is a free particle, and its wave function can be any linear combination of functions like that in Eq. (6-5). Since the particle cannot get outside the box, the wave function must be zero everywhere outside the box. Solutions to Schroedinger's equation must be *continuous*. That means that we must find wave functions for a free particle inside the box which go to zero at the two ends of the box, matching the wave function outside the box. If the two ends of the box are at $x = 0$ and $x = L$, then we require that the wave function $\psi(x, t)$ satisfy $\psi(0, t) = 0$ and $\psi(L, t) = 0$.

Consider the following wave function,

$$\psi = A\exp(ikx - i\omega t) - A\exp(-ikx - i\omega t). \qquad (6\text{-}25)$$

This is just a combination of two traveling waves, each going in opposite directions. Using Eq. (6-3), we may simplify this expression and obtain

$$\psi(x, t) = 2iA\sin(kx)\exp(-i\omega t). \qquad (6\text{-}26)$$

Problem 6-10. Derive Eq. (6-26) from Eq. (6-25).

We see that this wave function already satisfies the condition $\psi(0, t) = 0$ for any value of t. The other condition, $\psi(L, t) = 0$ may be satisfied by requiring that $kL = n\pi$, where n is any integer. This means that the wave number k must be restricted to values

$$k = n(\pi/L). \qquad (6\text{-}27)$$

Thus,

$$\psi = \begin{cases} 2iA\sin(n\pi x/L)\exp(-i\omega t), & 0 \le x \le L, \\ 0, & x \le 0 \text{ or } x \ge L. \end{cases} \quad (6\text{-}28)$$

Physically, only positive values of n need to be considered. Changing the sign of n only changes the sign of ψ and does not give us a physically different wave function. Also, if $n = 0$, then $\psi = 0$ everywhere, and there is no particle in the box. All the possible wave functions for a particle in the box thus correspond to non-zero values of k ($n \ne 0$) and thus non-zero momentum. A particle in a box *cannot* be at rest. This is in agreement with the Heisenberg uncertainty principle. The particle is confined to the box ($\Delta x = L$) and therefore its momentum is uncertain and cannot be exactly zero.

Problem 6-11. Consider an electron in a "box" which is 1.00 cm long. Using Eq. (6-27), find the smallest velocity the electron can have. Answer: 3.64 cm/s.

From Eq. (6-28), we find that the probability function $|\psi|^2$ of a particle in a box is given by

$$|\psi|^2 = 4A^2\sin^2(n\pi x/L). \quad (6\text{-}29)$$

Problem 6-12. Derive Eq. (6-29) from Eq. (6-28).

We plot $|\psi|^2$ in Fig. 6-4 for $n = 1, 2,$ and 3. We see in the figure that the probability of finding the particle is not distributed evenly across the length of the box. The particle spends more of its time in some regions of the box than in others. This is certainly not what classical physics would predict for a particle bouncing between two ends of a box. Classical physics would predict that the particle would spend an equal amount of time in every region of the box. The probability function should be a constant everywhere inside the box.

Quantum mechanics gives us a result very different from classical physics. For example, for $n = 1$ we see from the graph of $|\psi|^2$ that the electron spends most of its time near the center of the box but very little time near the edges. For $n = 2$, we

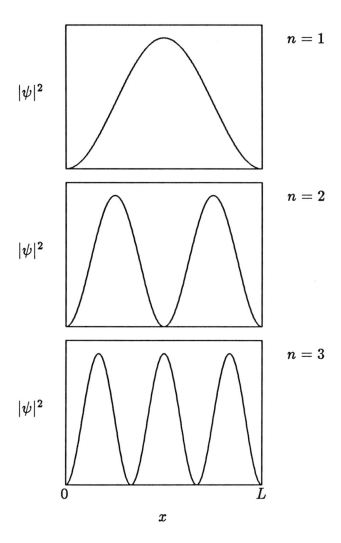

Fig. 6-4. The probability function of a particle in a box.

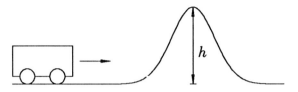

Fig. 6-5. Cart approaching a hill.

see that the electron spends most of its time near $x = \frac{1}{4}L$ and $x = \frac{3}{4}L$ but is *never* found at the center of the box ($|\psi|^2$ is zero at $x = \frac{1}{2}L$). How can the electron get from one end of the box to the other if it is never at the center? Such a question is classical in nature and reveals how our thinking is so very much influenced by the tenets of classical physics. In quantum mechanics, in order for a particle to move from one position to another, it is not necessary to require that it be found at every point in between.

6-5 Tunneling

Let us now examine one more non-classical phenomenon which we can predict with quantum mechanics. Consider a cart of mass m and kinetic energy E approaching a hill of height h (see Fig. 6-5).

The potential energy at the top of the hill is $U = mgh$ (with respect to the bottom of the hill). We call the hill a "potential energy barrier" of height U. From classical physics, we know that the cart will get over the top of the hill only if $E > U$.

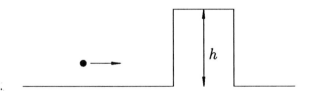

Fig. 6-6. Particle approaching a potential energy barrier.

Let us consider a similar situation using quantum mechanics. A free particle of kinetic energy $E = \hbar\omega = \hbar^2 k^2/2m$ approaches a potential energy barrier of height U which is *greater* than E (see Fig. 6-6). We consider the barrier to be "square" for the sake of simplicity. Let us find a solution to Schroedinger's equation *inside* the barrier. If we try a wave function of a free particle like Eq. (6-5), we obtain from Schroedinger's equation,

$$\hbar^2 k^2/2m + U = \hbar\omega. \tag{6-30}$$

Remember that $\hbar\omega$ is the *total* energy of the particle and is conserved, i.e., $\hbar\omega = E$ everywhere. Thus

$$k = \sqrt{\frac{2m(E-U)}{\hbar^2}}. \tag{6-31}$$

But $U > E$ and thus the quantity inside the square root is *negative*. The wave number k is an imaginary number. We can write k as

$$k = i\alpha, \tag{6-32}$$

where α is a real number given by

$$\alpha = \sqrt{\frac{2m(U-E)}{\hbar^2}}. \tag{6-33}$$

The wave function inside the potential barrier thus has the form

$$\psi = \exp(-\alpha x)\exp(-i\omega t), \tag{6-34}$$

and the probability function has the form

$$|\psi|^2 = A^2 \exp(-2\alpha x). \tag{6-35}$$

The probability of finding the particle inside the potential barrier decays exponentially to zero. Thus, the particle can penetrate the barrier to some extent but not very far. However, if the barrier is so narrow that the wave function does not decay very much inside the barrier, there is a finite probability that the particle can get through the barrier to the other side. This phenomenon is called **tunneling**. In order for the particle to have a reasonable probability of tunneling through a barrier, its width should not be more than a few times $1/\alpha$.

Tunneling has been observed experimentally. It has been found that if a small voltage is applied across a thin insulating layer between two pieces of metal, an electric current will flow due to electrons tunneling across the insulator. Careful measurements show that this current depends exponentially on the thickness of the insulating layer, as we would expect from Eq. (6-35). We will discuss other examples of tunneling in later chapters.

Problem 6-13. Consider an electron with kinetic energy equal to 5 eV. If it approaches a potential energy barrier of 6 eV, how narrow should the barrier be to allow the electron to tunnel through? Answer: less than approximately 2 Å.

6-6 Wave Functions in Three Dimensions

Up to this point, we have done everything in one-dimensional space. The mathematics is much simpler in one-dimensional space, and we will continue to use this approach throughout most of this book. However, the extension to three-dimensional space is not too difficult. The wave function ψ in three-dimensional space is a function of x, y, z, and t, and Schroedinger's equation is written as

$$-\frac{\hbar^2}{2m}\left(\frac{\partial^2\psi}{\partial x^2}+\frac{\partial^2\psi}{\partial y^2}+\frac{\partial^2\psi}{\partial z^2}\right)+U\psi=i\hbar\frac{\partial\psi}{\partial t}. \qquad (6\text{-}36)$$

For a free particle, $U = 0$, and the solution to Schroedinger's equation is a three-dimensional traveling wave,

$$\psi = A\exp(i\mathbf{k}\cdot\mathbf{r}-i\omega t), \qquad (6\text{-}37)$$

where
$$\mathbf{r} = x\hat{\mathbf{i}} + y\hat{\mathbf{j}} + z\hat{\mathbf{k}} \tag{6-38}$$
and **k** is the wave vector,
$$\mathbf{k} = k_x\hat{\mathbf{i}} + k_y\hat{\mathbf{j}} + k_z\hat{\mathbf{k}}. \tag{6-39}$$

The group velocity of the wave function is given by
$$\mathbf{v}_g = \frac{\partial \omega}{\partial k_x}\hat{\mathbf{i}} + \frac{\partial \omega}{\partial k_y}\hat{\mathbf{j}} + \frac{\partial \omega}{\partial k_z}\hat{\mathbf{k}}. \tag{6-40}$$

Problem 6-14. Show that Eq. (6-37) is a solution of Schroedinger's equation for $U = 0$.

Problem 6-15. Show that the group velocity given by Eq. (6-40) is equal to the actual velocity of the particle described by the wave function in Eq. (6-37). Use Eq. (6-21).

CHAPTER 7

FREE-ELECTRON QUANTUM MODEL OF METALS

7-1 Particle in a Box

In this chapter, we will modify the classical model we developed in Chapter 4. We will apply some quantum mechanics we introduced in Chapter 6 plus some additional concepts as we need them. As in the classical model, we will consider the conduction electrons to be free. We neglect their interaction with the positive ions and also with each other. We only require that the electrons remain within the metal. Thus, the electrons behave like particles trapped in a "box." We treated this problem in the last chapter for the one-dimensional case. Let us now extend that treatment to three dimensions.

Consider a cubic "box" with sides of length L. The wave function ψ inside the box is simply a solution of Schroedinger's equation for a free particle. Outside the box, ψ must be zero. If we put one corner of the box at the origin, as shown in Fig. 7-1, then we must require that the wave function $\psi(x,y,z,t)$ satisfy the following boundary conditions:

$$\begin{aligned}\psi(0,y,z,t) &= 0, \\ \psi(L,y,z,t) &= 0, \\ \psi(x,0,z,t) &= 0, \\ \psi(x,L,z,t) &= 0, \\ \psi(x,y,0,t) &= 0, \\ \psi(x,y,L,t) &= 0.\end{aligned} \quad (7\text{-}1)$$

Similar to Eq. (6-26), we try a wave function of the form,

$$\psi(x,y,z,t) = A\sin(k_x x)\sin(k_y y)\sin(k_z z)\exp(-i\omega t), \quad (7\text{-}2)$$

where k_x, k_y, k_z are components of the wave vector,

$$\mathbf{k} = k_x \hat{\mathbf{i}} + k_y \hat{\mathbf{j}} + k_z \hat{\mathbf{k}}. \quad (7\text{-}3)$$

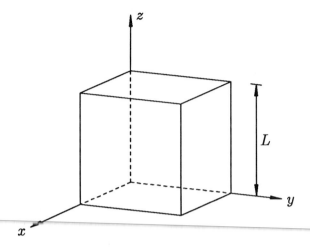

Fig. 7-1. Electrons in a box of side L.

Problem 7-1. Show that Eq. (7-2) is a solution of the three-dimensional Schroedinger equation for a free particle [Eq. (6-36) with $U = 0$].

If we require ψ to satisfy the boundary conditions of Eq. (7-1), we find that the components of **k** must be restricted to values,

$$k_x = n_x(\pi/L),$$
$$k_y = n_y(\pi/L), \qquad (7\text{-}4)$$
$$k_z = n_z(\pi/L),$$

where n_x, n_y, n_z are integers. As in the one-dimensional case, we consider only positive non-zero values of n_x, n_y, and n_z.

Problem 7-2. Show that the wave function ψ given by Eq. (7-2) satisfies the boundary conditions given by Eq. (7-1) if the values of k_x, k_y, and k_z are restricted to those of Eq. (7-4).

7-2 Periodic Boundary Conditions

There is another way to treat free conduction electrons in a metal which is actually preferable to "particles in a box." The cubic box of side L has surfaces. We remove the surfaces by stacking identical cubes of side L in a simple-cubic arrangement so that they fill all space as in Fig. 1-6. The cubes are repeated throughout all space. Each wave function also fills all space. Since we are setting up a model for electrons within a single cube of metal, we want *everything* about each cube to be identical. (Every cube represents the actual cube of metal which we are considering.) Therefore, each wave function must be identical in every cube. The boundary conditions of Eq. (7-1) are thus replaced with

$$\psi(x+L,y,z,t) = \psi(x,y,z,t),$$
$$\psi(x,y+L,z,t) = \psi(x,y,z,t), \qquad (7\text{-}5)$$
$$\psi(x,y,z+L,t) = \psi(x,y,z,t).$$

These are called **periodic boundary conditions**. We find that in this case, the wave function for a free electron given by Eq. (6-37),

$$\psi(x,y,z,t) = A\exp(ik_x x + ik_y y + ik_z z - i\omega t), \qquad (7\text{-}6)$$

can satisfy the periodic boundary conditions of Eq. (7-5) if we restrict the components of **k** to the values,

$$k_x = n_x(2\pi/L),$$
$$k_y = n_y(2\pi/L), \qquad (7\text{-}7)$$
$$k_z = n_z(2\pi/L).$$

Problem 7-3. Show that the wave function ψ given by Eq. (7-6) satisfies the boundary conditions given by Eq. (7-5) if the values of k_x, k_y, and k_z are restricted to those of Eq. (7-7).

In this case, n_x, n_y, n_z may take on negative values as well as positive values since these give us *different* wave functions ψ. This is a principal advantage of using periodic boundary conditions in place of those in Eq. (7-1). The wave vector **k** can now point in negative as well as positive directions, and we can now associate **k** with the direction of the electron's velocity. Using Eq. (7-7), we find that the allowed wave vectors are

$$\mathbf{k} = (2\pi/L)(n_x\hat{\imath} + n_y\hat{\jmath} + n_z\hat{k}). \qquad (7\text{-}8)$$

Each allowed **k** corresponds to a different wave function ψ of an electron.

7-3 Density of States

If we plot all the allowed wave vectors in **k**-space (reciprocal space), we obtain a set of points which forms an sc lattice as shown in Fig. 7-2. The distance between adjacent "lattice" points is $2\pi/L$. Remember that for an sc lattice of lattice parameter a, the density of lattice points is given by a^{-3}. In the present case, we have an sc lattice in **k**-space with lattice parameter $2\pi/L$. Therefore, the density of these points in **k**-space is $(2\pi/L)^{-3} = V/(2\pi)^3$, where $V = L^3$, the volume of the cube.

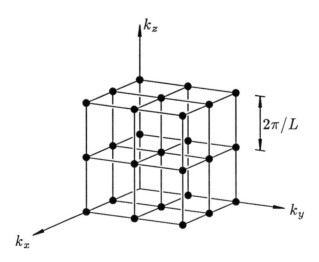

Fig. 7-2. Allowed wave vectors of conduction electrons in a metal.

Before we proceed further here, we must introduce another property of electrons. An electron spins about an axis. Since it has charge, the spinning motion gives rise to a magnetic moment (like current in a loop of wire). It has been found that the direction of this magnetic moment due to spinning is quantized and can take on only one of two possible values. These are called the **spin states** of the electron. For every allowed wave function ψ of an electron, there are two *electron states*, corresponding to the two possible directions which the magnetic moment of the electron can point.

Every allowed value of **k** given by Eq. (7-8) corresponds to a different wave function and thus represents *two* electron states. If the density of these allowed wave vectors **k** in k-space is equal to $V/(2\pi)^3$, then the density of electron states in k-space is equal to twice that value, or $2V/(2\pi)^3$. Denoting the density of electron states in reciprocal space by $g(\mathbf{k})$, we write

$$g(\mathbf{k}) = 2V/(2\pi)^3. \tag{7-9}$$

The number of electron states within some volume V_k in k-space is given by the product $V_k\, g(\mathbf{k})$.

Let us next calculate the number of electron states with energy less than some value E. For free electrons,

$$E = p^2/2m = \hbar^2 k^2/2m. \tag{7-10}$$

Solving for k,

$$k = \sqrt{2mE/\hbar^2}. \tag{7-11}$$

Thus, the number of electron states with energy less than E is the same as the number of states with wave number less than the value of k given by Eq. (7-11). These states are contained within a sphere of radius k in k-space. The volume of this sphere is given by

$$V_k = \tfrac{4}{3}\pi k^3 = \tfrac{4}{3}\pi(2mE/\hbar^2)^{3/2}. \tag{7-12}$$

To obtain the number of states $N(E)$ within this sphere, we multiply V_k by the density of states in k-space given by

Eq. (7-9) and obtain

$$N(E) = V_k \, g(\mathbf{k}) = (V/3\pi^2)(2mE/\hbar^2)^{3/2}. \qquad (7\text{-}13)$$

Problem 7-4. Consider what would happen if instead of using periodic boundary conditions, we retained the boundary conditions of Eq. (7-1) for particles in a box. Find the resulting density of states $g(\mathbf{k})$ in k-space. Show that the number of states $N(E)$ with energy less than some value E is the same as Eq. (7-13).

We next obtain the **density of states** $g(E)$ as a function of E. By density of states $g(E)$, we mean that the number of electron states between two energies, E_1 and E_2, is given by by the integral of $g(E)$ from E_1 to E_2. The quantity $N(E)$ in Eq. (7-13) above is the number of states between energies 0 and E. Thus,

$$N(E) = \int_0^E g(E) \, dE, \qquad (7\text{-}14)$$

or

$$g(E) = \frac{dN(E)}{dE}, \qquad (7\text{-}15)$$

which, using Eq. (7-13), gives us

$$g(E) = (V/2\pi^2)(2m/\hbar^2)^{3/2} E^{1/2}. \qquad (7\text{-}16)$$

A plot of $g(E)$ is shown in Fig. 7-3. It is simply a parabola on its side. As stated above, by definition, the number of electron states between two energies, E_1 and E_2, is given by the area under the curve between E_1 and E_2. This is shown by the shaded area in Fig. 7-3.

7-4 Pauli's Exclusion Principle

So far, we have only found the possible *states* of an electron in the metal. If we have N conduction electrons, we now must

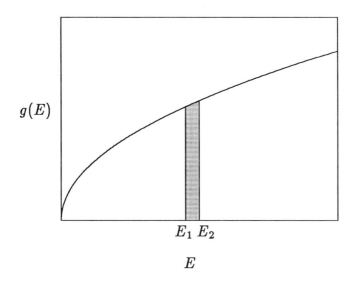

Fig. 7-3. The density of states $g(E)$. The number of electron states between E_1 and E_2 is given by the shaded area under the curve.

consider which states we will find these electrons in. Let us first consider the case of absolute zero temperature ($T = 0$). None of the electrons have any thermal energy and must therefore occupy the states of lowest energy available. We might expect all of the electrons to be in the same state (the state of lowest energy). However, this cannot happen because of another rule in quantum mechanics called **Pauli's exclusion principle.**

Pauli's exclusion principle states that no more than one electron can be in the same state at the same time. This is simply a fundamental principle of quantum mechanics which is analogous to the classical concept that two objects cannot be at the same place at the same time. However, Pauli's exclusion principle only applies to a class of particles called **fermions**. (Electrons, protons, and neutrons are examples of fermions.) All other particles (called **bosons**) do not need to obey this principle. The helium atom, which of course is a composite of fermions, happens to be a boson, for example.

Pauli's exclusion principle has dramatic consequences for

conduction electrons in a metal. At $T = 0$, each electron tries to occupy the lowest available energy state. But since they cannot all be in the same state, some electrons must by necessity occupy states of higher energy. The situation is analogous to filling a glass with water. Because of gravity, all the water molecules would like to be at the bottom of the glass, but because of a "volume exclusion principle," only a small fraction are at the bottom. In fact, some water molecules are forced to be quite far removed from the bottom, depending on how much water we pour into the glass.

Similarly, the energy of the highest state occupied by an electron in a metal depends on how many conduction electrons there are. With all the conduction electrons occupying states, the energy of the highest occupied state is called the **Fermi energy** E_F. The number of states $N(E_F)$ with energy between 0 and E_F must therefore be equal to the total number of conduction electrons N. From Eq. (7-13), we thus have

$$N = V_k\, g(\mathbf{k}) = (V/3\pi^2)(2mE_F/\hbar^2)^{3/2}. \qquad (7\text{-}17)$$

Solving for E_F, we obtain

$$E_F = (\hbar^2/2m)(3\pi^2 n)^{2/3}, \qquad (7\text{-}18)$$

where $n = N/V$, the density of conduction electrons in the metal. Note that E_F does not depend on the volume V of the metal. The velocity of the electrons at the Fermi energy is called the **Fermi velocity** v_F.

Problem 7-5. Find the Fermi energy and Fermi velocity of the conduction electrons in sodium metal (Na). Answer: 3.1 eV, 1.05×10^6 m/s.

Conduction electrons fill all the states from zero energy up to the Fermi energy, and thus there is a wide distribution of velocities among the electrons, from very small velocity up to the Fermi velocity. As seen in the above problem, the Fermi velocity is typically of the order of 10^6 m/s. Remember that this is at $T = 0$ where classical physics demands that *none* of the conduction electrons have any kinetic energy at all. Even at room temperature, the Fermi velocity is much greater than the typical velocity an electron may attain from thermal energy alone.

Since the area under the $g(E)$ curve gives us the number of electron states, we can represent diagrammatically the occupation of these states by the area under the curve up to $E = E_F$, as shown in Fig. 7-4. At temperatures $T > 0$, there is some thermal energy available to each of the conduction electrons. This thermal energy is usually much less than E_F. Thus, most of the electrons cannot acquire any of this thermal energy available to them because the states slightly higher in energy than their own are already occupied by other electrons and Pauli's exclusion principle will not allow them to share the energy level with another electron. However, electrons near the Fermi level *can* be thermally excited since there are many un-

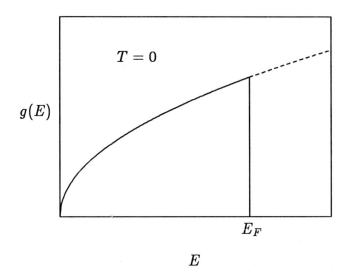

Fig. 7-4. Occupation of states by conduction electrons at $T = 0$.

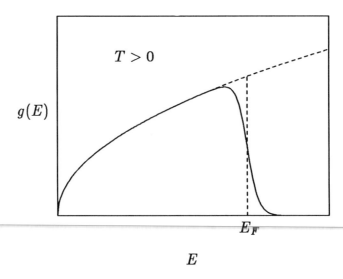

Fig. 7-5. Occupation of states by conduction electrons at $T > 0$.

occupied states above E_F for them to go into. We thus get an occupation of states like that shown in Fig. 7-5.

7-5 Fermi-Dirac Distribution Function

Let us find a quantitative expression for the curve in Fig. 7-5. Consider two electron states. We label these two states 1 and 2. They correspond to some energy levels E_1 and E_2. An electron in state 1 has energy E_1, and an electron in state 2 has energy E_2. We choose the labels of the two states such that $E_2 > E_1$.

Let us assume that at some time t there is an electron in state 1 and *no* electron in state 2. Let W_{12} be the probability that the electron in state 1 will make a transition to state 2 during some short time interval Δt. Similarly, let us assume that at some other time t there is an electron in state 2 and *no* electron in state 1. Let W_{21} be the probability that the electron in state 2 will make a transition to state 1 during Δt.

The probabilities W_{12} and W_{21} have a very simple quantitative relation. We will state this relation without proof and

CHAPTER 7   FREE-ELECTRON QUANTUM MODEL OF METALS   143

use it as a starting point for our derivation:

$$W_{21} = W_{12} \exp\left(\frac{E_2 - E_1}{k_B T}\right), \tag{7-19}$$

where T is the temperature and k_B is called **Boltzmann's constant**. This equation states that W_{21} is greater than W_{12} (remembering that we chose $E_2 > E_1$). An electron is more likely to go from a state of high energy (state 2) to a state of low energy (state 1) than vice versa. Furthermore, at lower temperature, the difference between these two probabilities is even greater. Electrons have less thermal energy and thus find it more difficult to go from state 1 to state 2.

Pauli's exclusion principle tells us that each state in the metal can at most be occupied by one electron. No two electrons can occupy the same state simultaneously. This restriction has serious consequences for the distribution of electrons among the states at thermal equilibrium.

Remember that we said W_{12} was the probability of an electron going from 1 to 2, assuming that there was an electron in state 1 and *no* electron in state 2. If there were an electron already in state 2, then, of course, the probability of the electron going from 1 to 2 is zero. Two electrons cannot be in state 2 simultaneously.

So, let P_1 be the probability that an electron is in state 1 and P_2 be the probability that an electron is in state 2. The probability that *no* electron is in state 1 is $1 - P_1$, and the probability that *no* electron is in state 2 is $1 - P_2$. The probability W'_{12} that *some* electron will go from 1 to 2 during the time interval Δt is equal to W_{12} times the probability that there is an electron in state 1 (no transition can take place unless an electron is there to make the transition) times the probability that *no* electron is in state 2:

$$W'_{12} = W_{12} P_1 (1 - P_2). \tag{7-20}$$

Similarly, the probability W'_{21} that *some* electron will go from 2 to 1 is given by

$$W'_{21} = W_{21} P_2 (1 - P_1). \tag{7-21}$$

At equilibrium, the average occupation P_n of any state is constant in time (by definition of equilibrium). Thus, the probability W'_{12} that some electron will go from 1 to 2 must be equal to the probability W'_{21} that some electron will go from 2 to 1. Over a given time interval, an equal number of electrons must go from 1 to 2 and from 2 to 1. Otherwise, the average occupation of these states would change, and the electrons would not be at equilibrium. This requires that Eq. (7-20) be equal to Eq. (7-21):

$$W_{12}P_1(1 - P_2) = W_{21}P_2(1 - P_1). \tag{7-22}$$

Combining this with Eq. (7-19) and rearranging the terms, we obtain

$$\frac{P_1}{1-P_1}\exp(E_1/k_BT) = \frac{P_2}{1-P_2}\exp(E_2/k_BT). \tag{7-23}$$

Since states 1 and 2 were chosen arbitrarily, this relation must hold for *any* pair of states in the metal. Thus,

$$\frac{P_n}{1-P_n}\exp(E_n/k_BT) = \text{constant} \tag{7-24}$$

for any electron state n in the metal.

Consider a state in the metal such that $P_n = \frac{1}{2}$. (The probability that an electron would be in that state is $\frac{1}{2}$.) We will define the energy level corresponding to that state to be the Fermi energy E_F. (There may or may not be an actual electron state at E_F. But we can always find E_F such that if there *were* a state at E_F, then P_n would be $\frac{1}{2}$.) Substituting $P_n = \frac{1}{2}$ and $E_n = E_F$ into Eq. (7-24), we obtain

$$\text{constant} = \exp(E_F/k_BT). \tag{7-25}$$

Putting this into Eq. (7-24) and solving for P_n, we obtain

$$P_n = \frac{1}{\exp\left(\frac{E_n-E_F}{k_BT}\right)+1}. \tag{7-26}$$

CHAPTER 7 FREE-ELECTRON QUANTUM MODEL OF METALS

We can remove the index n from this expression and simply write that the probability [which we denote by $f_D(E)$] that a state of energy E is occupied by an electron is given by

$$f_D(E) = \frac{1}{\exp\left(\frac{E-E_F}{k_B T}\right) + 1}. \tag{7-27}$$

This is called the **Fermi-Dirac distribution function**.

At $T = 0$, $f_D(E)$ has a very simple form.

$$f_D(E) = \begin{cases} 1, & E < E_F, \\ 0, & E > E_F. \end{cases} \tag{7-28}$$

This means that the probability that a state below the Fermi level is occupied is equal to 1, and the probability that a state above the Fermi level is occupied is equal to 0. All states below E_F are occupied and all states above E_F are unoccupied.

The Fermi-Dirac distribution function is shown in Fig. 7-6 for various temperatures T. We see that at higher temperatures, more electrons are excited above the Fermi level. Note that in all cases, $f_D(E_F) = \frac{1}{2}$.

Problem 7-6. Show that at $T = 0$, the function $f_D(E)$ in Eq. (7-27) has the form given by Eq. (7-28).

Problem 7-7. How far below (or above) the Fermi level will we find states which are 90% occupied by electrons at 300 K? Repeat this calculation for 99%, 10%, and 1%. Answer: 0.057 eV below, 0.12 eV below, 0.057 eV above, 0.12 eV above.

Problem 7-8. At what temperatures would the graphs in Fig. 7-6 represent the probability of occupation of states in sodium metal (Na)? You may use the result of Problem 7-5. Answer: 1090 K, 2180 K.

The density of *occupied* electron states in a metal is given by the product $f_D(E)g(E)$ since $g(E)$ is the density of states

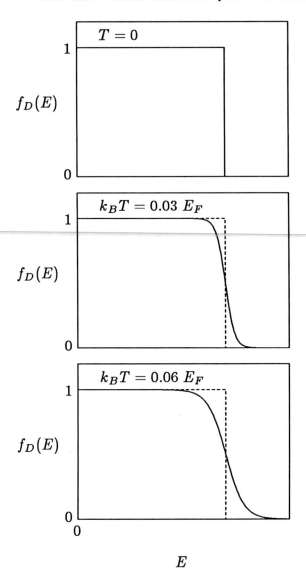

Fig. 7-6. The Fermi-Dirac distribution function at $T = 0$ and at two different values of $T > 0$.

CHAPTER 7 FREE-ELECTRON QUANTUM MODEL OF METALS

available for occupation by electrons and $f_D(E)$ is the probability that those states will be occupied. This is what we plotted in Figs. 7-4 and 7-5. The total number of conduction electrons is given by

$$N = \int_0^\infty f_D(E)g(E)\,dE. \tag{7-29}$$

The Fermi energy must be chosen to satisfy this equation. At $T = 0$, we simply obtain Eq. (7-18). At $T > 0$, the integral in Eq. (7-29) cannot be done analytically. From numerical integration, one finds that the value of E_F decreases slightly with increasing T. We will neglect this temperature variation in E_F and consider E_F to be given by Eq. (7-18) at all temperatures.

Problem 7-9. Obtain Eq. (7-18) from Eq. (7-29) for $T = 0$.

Problem 7-10. Why does E_F decrease with increasing T?

7-6 Electrical Conductivity

We will now examine electrical conductivity using the free-electron quantum model we have developed here. At $T = 0$, the occupied electron states all fall within a sphere in k-space (see Fig. 7-7). The energy of the states on the surface of this sphere is equal to E_F. All states inside this sphere have energies less than E_F and are occupied. All states outside this sphere have energies greater than E_F and are unoccupied. The surface of this sphere in k-space is called the **Fermi surface**. For $T > 0$, the picture is essentially the same except that some states just outside the Fermi surface are occupied and some states just inside the Fermi surface are unoccupied.

Note that this sphere is centered on $\mathbf{k} = 0$, and thus for every state at some $\mathbf{k}$ inside the sphere, there is another state at $-\mathbf{k}$ inside the sphere. Consequently, for every electron with momentum $\mathbf{p} = \hbar\mathbf{k}$, there is another electron with momentum $\mathbf{p}' = -\hbar\mathbf{k}$. We see then that the average momentum $\langle \mathbf{p} \rangle$ of all

148   CHAPTER 7   FREE-ELECTRON QUANTUM MODEL OF METALS

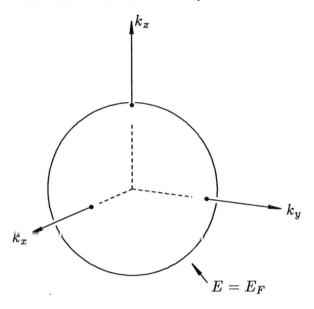

Fig. 7-7. The Fermi surface in k-space.

the electrons is zero since for every **p** there is a −**p** to cancel it. Also, from **p** = m**v**, we see that ⟨**v**⟩ = 0, and therefore the electrons carry no net current through the metal, as we would expect.

In the presence of an external electric field $\mathcal{E}$, we saw in Chapter 4 using the classical model that each electron acquires an average non-zero velocity, called the drift velocity $\mathbf{v}_d$. We can do the same thing here by giving each electron a velocity $\mathbf{v}_d$ in addition to the velocity already associated with its original state at **k**. Changing the velocity of an electron also changes its wave vector **k**. The amount of change in **k** is given by

$$\Delta \mathbf{k} = (1/\hbar)\Delta \mathbf{p} = (m/\hbar)\mathbf{v}_d. \tag{7-30}$$

Thus, all electrons move to new states, displaced by Δ**k** from their old states. The occupied states still fall within a sphere, but now it is a sphere centered at **k** = Δ**k** as shown in Fig. 7-8. Note that Δ**k** is in a direction opposite to that of $\mathcal{E}$ since the force on the electrons due to $\mathcal{E}$ is in a direction opposite to $\mathcal{E}$ because of the negative charge of the electrons.

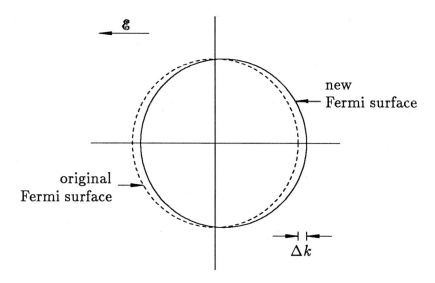

Fig. 7-8. Fermi surface displaced in presence of electric field $\mathcal{E}$.

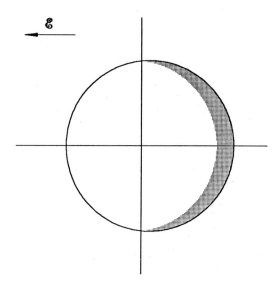

Fig. 7-9. Uncompensated electron states (the shaded area) in the presence of an electric field $\mathcal{E}$.

We can see here that now not all electrons can be put into **k** and −**k** pairs. Some states on the left side of the sphere in Fig. 7-8 that used to be occupied are now unoccupied, and some states on the right side that used to be unoccupied are now occupied. There is an imbalance, and $\langle p \rangle$ is no longer zero. We thus get an electric current.

Since $v_d \ll v_F$, the effect is small, and most electrons are still compensated, that is, they come in **k** and −**k** pairs. The uncompensated electrons are found in a small region near the Fermi surface on the right side, as shown in Fig. 7-9. All the rest of the electrons are compensated and their average momentum $\langle p \rangle$ is zero. Thus they do not contribute to the electric current. The electric current arises from the *uncompensated* electrons near the Fermi surface. This is only a small fraction of the total number of conduction electrons.

Thus we see that the quantum picture of electrical conductivity is quite different from the classical picture. In the classical picture, all the conduction electrons participate in the electric current by acquiring an average velocity v_d. In the quantum picture, only a small fraction of the conduction electrons participate in the current, and they have a velocity approximately equal to the Fermi velocity v_F which is many orders of magnitude larger than v_d.

Problem 7-11. Find the radius of the Fermi surface in sodium metal (Na). If an electric field E=1.0 V/m is applied, causing current to flow, find the drift velocity of the conduction electrons and the displacement Δk of the Fermi surface. Answer: 9.07×10^9 m^{-1}, 5.90×10^{-3} m/s, 51.0 m^{-1}.

To calculate the electrical conductivity, we proceed in much the same manner as we did for the classical model. The fraction of the conduction electrons which are uncompensated is of the order of v_d/v_F. Thus the density of these electrons is approximately equal to $n(v_d/v_F)$, where n is the density of *all* the conduction electrons. In Chapter 4 [see Eq. (4-7)], we found

CHAPTER 7   FREE-ELECTRON QUANTUM MODEL OF METALS   151

that the current density J is equal to the density of conducting electrons (nv_d/v_F here) times the charge e of an electron times the velocity of the conducting electrons (v_F here). This gives us

$$J \cong (nv_d/v_F)ev_F = nev_d, \qquad (7\text{-}31)$$

which is the same as Eq. (4-7) in the classical case. In the quantum model, the meaning of v_d is the displacement of the Fermi surface, as given by Eq. (7-30), and not an average "drift velocity," which was the original meaning of v_d in the classical model.

This displacement of the Fermi surface is governed by how far the electrons near the Fermi surface can be accelerated by the electric field before they are scattered by some kind of collision. This acceleration pushes these electrons into new energy states outside the original Fermi surface. All the other electrons inside the Fermi surface are also accelerated by the field, but they cannot go any further than the electrons near the Fermi surface, due to Pauli's exclusion principle. The electrons near the Fermi surface are in the way, occupying states that the other electrons might otherwise be accelerated into. Thus we see that the displacement $\Delta \mathbf{k}$ is limited by the scattering of the electrons near the Fermi surface.

Using the same reasoning as we did for the classical model in Chapter 4, we can easily show that the extra velocity picked up by the electrons near the Fermi surface is given by

$$v_d = e\tau_F \mathscr{E}/m, \qquad (7\text{-}32)$$

as in Eq. (4-6), except that now τ_F is the average time between collisions of electrons near the Fermi surface. Putting this into Eq. (7-31) and remembering that electrical conductivity is defined by the relation $J = \sigma \mathscr{E}$ [Eq. (4-9)], we have, similar to Eq. (4-10),

$$\sigma = ne^2 \tau_F/m. \qquad (7\text{-}33)$$

From experimental values of σ, we can calculate τ_F. These are the same as the values of τ which we calculated for the classical model in Chapter 4.

The average distance ℓ which one of these electrons travels between collisions is given by

$$\ell = v_F \tau_F. \qquad (7\text{-}34)$$

Problem 7-12. Find the average distance ℓ which a conduction electron near the Fermi surface travels between collisions in sodium metal (Na). You may use the results of Problems 4-4 and 7-5. Answer: 352 Å.

As can be seen from the above problem, an electron in Na travels a rather large distance between collisions, considering that the distance between Na ions is only about 4 Å. How can an electron go so far without colliding with a Na ion? We will answer this question later, after we have developed our model of electrical conductivity in metals a bit further.

CHAPTER 8

BAND THEORY OF METALS

8-1 Interaction with Ions

In the last chapter, we considered the conduction electrons in a metal to be free. We neglected the interaction of the electrons with the positive ions and with other electrons. In this chapter, we will see what happens when we include the interaction of the electrons with the positive ions. We will do much of the treatment in one dimension since it is easier to visualize.

Consider a one-dimensional metal. The positive ions are separated by a distance a, the lattice parameter. In the free-electron model, the energy E of a conduction electron with wave vector **k** is given by Eq. (7-10),

$$E = (\hbar^2/2m)k^2. \tag{8-1}$$

The plot of E as a function of k is a parabola, as shown in Fig. 8-1.

The interaction with the positive ions affects this relationship between E and k. It is easiest to see this at the boundary of the first Brillouin zone, $k = \pm\pi/a$. There, two free-electron wave functions are

$$\psi = A\exp(\pm i\pi x/a - i\omega t). \tag{8-2}$$

One of these wave functions represents an electron traveling in the $+x$ direction, and the other represents an electron traveling in the $-x$ direction. At $k = \pm\pi/a$, the de Broglie wavelength of the electron satisfies Bragg's law [Eq. (2-10)] for diffraction.

Problem 8-1. Show that an electron with wave number $k = \pi/a$ in a one-dimensional lattice satisfies Bragg's law.

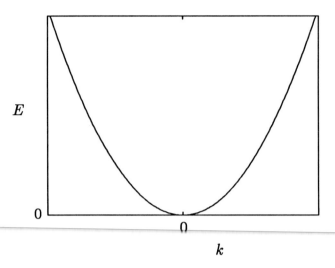

Fig. 8-1. The energy of a free electron.

An electron which satisfies Bragg's law is reflected by the lattice. As a result, the electron cannot travel through the lattice since it is continually being reflected, first in one direction and then the other. Consequently, the wave functions in Eq. (8-2) are inappropriate for describing the electron states at the first Brillouin zone boundary, since those wave functions describe *traveling* electrons.

We can form appropriate wave functions, ψ_1 and ψ_2, which do describe the electron states at the first Brillouin zone boundary by taking linear combinations of the two wave functions in Eq. (8-2). ψ_1 is the sum of those wave functions, and ψ_2 is the difference. They are

$$\psi_1 = 2A\cos(\pi x/a)\exp(-i\omega t)$$

and

$$\psi_2 = 2iA\sin(\pi x/a)\exp(-i\omega t). \qquad (8\text{-}3)$$

Problem 8-2. Show that ψ_1 and ψ_2 in Eq. (8-3) are the sum and difference, respectively, of the two wave functions in Eq. (8-2).

The probability functions of these two electron states are given by
$$|\psi_1|^2 = 4A^2 \cos^2(\pi x/a)$$
and
$$|\psi_1|^2 = 4A^2 \sin^2(\pi x/a). \tag{8-4}$$
When we plot these two functions (see Fig. 8-2), we see that they are identical except for their position with respect to the positive ions. $|\psi_1|^2$ is maximum at the ions, and $|\psi_2|^2$ is zero at the ions. This means that an electron with the wave function ψ_1 spends more time near the positive ions than an electron with the wave function ψ_2.

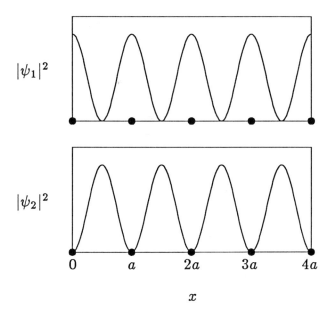

Fig. 8-2. Probability functions of ψ_1 and ψ_2. The black dots represent the positions of positive ions.

The potential energy U between the electron and the positive ion is negative (they attract each other) and increases in magnitude as they get closer together. Since an electron with wave function ψ_1 spends more time near the ions, its potential energy is more negative. Therefore, the total energy E (which is the sum of the kinetic and potential energy) of the electron with the wave function ψ_1 is *lower* than that of the electron with the wave function ψ_2. The state ψ_1 has a lower energy than the state ψ_2. In the free-electron model, both states have the *same* energy. The presence of the positive ions, though, "splits" the energy of those two states.

If we solve Schroedinger's equation, using a proper potential energy U of an electron in a one-dimensional metal, we obtain a relationship between E and k like that shown in Fig. 8-3. It appears very much like that of a free electron (Fig. 8-1), except for energy splittings at $k = \pm \pi/a$, $\pm 2\pi/a$, etc.

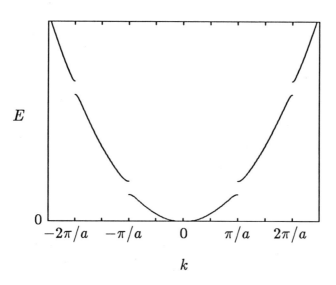

Fig. 8-3. Energy of an electron in a one-dimensional metal.

8-2 Bloch Functions

In solving Schroedinger's equation, we also find that the wave function ψ is affected by the interaction with the positive ions. It can be shown that, in general, the solution to Schroedinger's equation for electrons in any crystal can always be written in the form,

$$\psi = u_k(x) \exp(ikx - i\omega t), \qquad (8\text{-}5)$$

where $u_k(x)$ is a function which is periodic in a and repeats itself in each unit cell of the lattice, that is,

$$u_k(x + a) = u_k(x). \qquad (8\text{-}6)$$

The function $u_k(x)$ may be quite complicated. But whatever it is, it must be identical in every unit cell.

The wave function in Eq. (8-5) is called a **Bloch function**. We see from Eq. (8-5) that every wave function ψ is associated with some free-electron wave function $\exp(ikx - i\omega t)$. We can thus label every electron state with a wave number k, even for electrons which are no longer free. This is the meaning of k in Fig. 8-3. Also, the function $u_k(x)$ is generally different for each value of k.

Consider next a wave function labeled with some wave number k' *outside* the first Brillouin zone:

$$\psi = u_{k'}(x) \exp(ik'x - i\omega t). \qquad (8\text{-}7)$$

Remember that the reciprocal lattice vectors in the one-dimensional lattice are given by $G = 2\pi n/a$, where n is an integer. (See the discussion following Problem 3-5.) We know we can always find some reciprocal lattice vector G such that $k = k' + G$ is a wave vector *inside* the first Brillouin zone. If we put $k' = k - G$ into Eq. (8-7), we find that we can write ψ as

$$\psi = u_{n,k}(x) \exp(ikx - i\omega t), \qquad (8\text{-}8)$$

where

$$u_{n,k}(x) = u_{k'}(x) \exp(-i2\pi nx/a). \qquad (8\text{-}9)$$

Problem 8-3. Show that Eq. (8-7) can be written in the form of Eq. (8-8) if we substitute $k' = k - G$, where $G = 2\pi n/a$.

Problem 8-4. Show that the function $u_{n,k}(x)$ in Eq. (8-9) is periodic in a.

Since $u_{n,k}(x)$ is periodic in a (see above problem), we see that Eq. (8-8) is a Bloch function with a wave vector k *inside* the first Brillouin zone. In general, any Bloch function outside the first Brillouin zone can be rewritten as a Bloch function inside the first Brillouin zone. Thus, all electron states can be labeled with wave vectors k inside the first Brillouin zone. The function $u_{n,k}(x)$ is generally different from the function $u_k(x)$ which belongs to the Bloch function that was already in the first Brillouin zone. That is why we added the subscript n, to distinguish between them.

Since all electron states can be labeled with wave vectors k inside the first Brillouin zone, we can redraw Fig. 8-3 as shown in Fig. 8-4. For a given k, the different electron states with different energies are distinguished with the label $n = 1, 2, 3$, etc., as shown in Fig. 8-4. We see in the figure that at certain energies, no electron states exist. We have **bands** of allowed energies, separated by **gaps** of forbidden energies, as shown in Fig. 8-5. Each band is associated with one of the integers $n = 1, 2, 3$, etc. The labeling n, k of a Bloch function as in Eq. (8-8) tells us not only the wave number k (the horizontal axis in Fig. 8-5) but also which energy band n the electron state belongs to.

8-3 Three-Dimensional Crystals

Extension to three dimensions is not too difficult. Solutions to Schroedinger's equation give us Bloch functions of the form,

$$\psi = u_\mathbf{k}(\mathbf{r}) \exp(i\mathbf{k} \cdot \mathbf{r} - i\omega t), \qquad (8\text{-}10)$$

where

$$\mathbf{r} = x\hat{\mathbf{i}} + y\hat{\mathbf{j}} + z\hat{\mathbf{k}} \qquad (8\text{-}11)$$

CHAPTER 8 BAND THEORY OF METALS   159

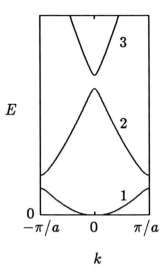

Fig. 8-4. Energy of electrons in the first Brillouin zone.

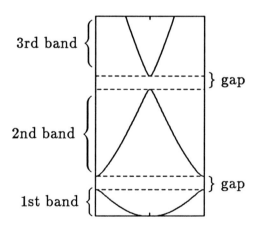

Fig. 8-5. Energy bands and gaps.

and
$$\mathbf{k} = k_x\hat{\mathbf{i}} + k_y\hat{\mathbf{j}} + k_z\hat{\mathbf{k}}. \tag{8-12}$$

The function $u_\mathbf{k}(\mathbf{r})$ is periodic in $\mathbf{R}$, the lattice vector, that is,
$$u_\mathbf{k}(\mathbf{r} + \mathbf{R}) = u_\mathbf{k}(\mathbf{r}), \tag{8-13}$$

and it thus repeats itself in every unit cell.

Any Bloch function for $\mathbf{k}$ outside the first Brillouin zone can be rewritten as a Bloch function inside the first Brillouin

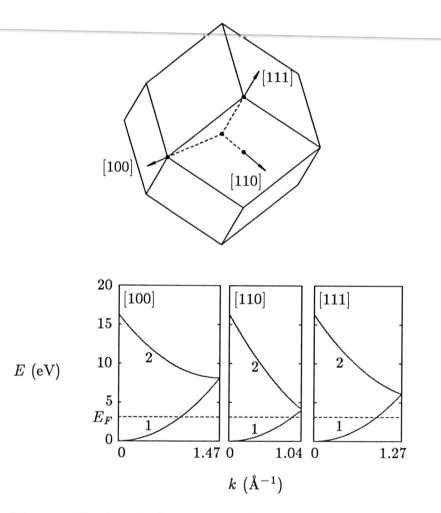

Fig. 8-6. The first Brillouin zone and the first two energy bands in sodium metal.

zone. This allows us to consider all the electron states to be inside the first Brillouin zone. If we plot the energy of these states from the center of the first Brillouin zone along some direction toward the boundary, we find a result similar to Fig. 8-5, that is, bands of energy separated by gaps.

In Fig. 8-6 we show the first two energy bands of sodium metal (Na) plotted along three different directions. The energies shown here are almost identical to those of free electrons. Along the [110] direction, we observe a gap between the two bands as in Fig. 8-5. Along the [100] and [111] directions, there happens to be no gap between these two bands. If we plot the density of states $g(E)$ for the first energy band, we get something that looks like Fig. 8-7. Near the bottom of the band ($E = 0$), the density of states $g(E)$ looks like the free-electron model (dashed line) in Fig. 7-3. At the top of the band, $g(E)$ goes to zero, as it must. The density of states in the second energy band would look somewhat similar, but would be at a higher energy. The third band, fourth band, etc., would be at higher energies yet. Some bands may overlap, and some bands

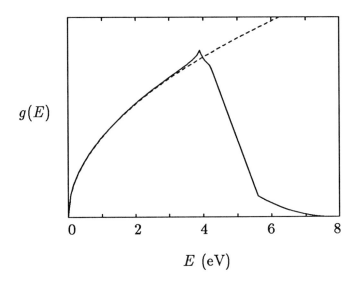

Fig. 8-7. Density of states for the first energy band in sodium metal. The dashed line is for the free-electron model.

may be separated by a gap. We will have more to say about this later.

Problem 8-5. Why is there a "spike" in $g(E)$ in Fig. 8-7 at about $E = 3.9$ eV? (Hint: examine Fig. 8-6 closely.)

8-4 Number of Electron States in a Band

Let us next determine the number of electron states in an energy band. Each energy band contains states throughout the entire first Brillouin zone. We found for the free-electron model that the density of states in k-space is given by $g(\mathbf{k}) = 2V/(2\pi)^3$. This relation still holds here since the Bloch functions must satisfy the same periodic boundary conditions given in Eq. (7-7). We must be careful to note, however, that $g(\mathbf{k})$ now means the density of states in k-space for a *single* energy band.

To find the number of states, we must first find the volume V_k of the first Brillouin zone. This is simply the volume of a primitive unit cell of the reciprocal lattice. The answer is simple. If V_c is the volume of a primitive unit cell of the direct lattice, then the volume V_k of a primitive unit cell of the reciprocal lattice is given by

$$V_k = (2\pi)^3/V_c. \tag{8-14}$$

This holds true for *any* Bravais lattice.

We will demonstrate the validity of Eq. (8-14) for the sc lattice. If the lattice parameter of the direct lattice is a, then the lattice parameter of the reciprocal lattice is $2\pi/a$. Thus, $V_c = a^3$, and $V_k = (2\pi/a)^3$, in agreement with Eq. (8-14).

Problem 8-6. Demonstrate that Eq. (8-14) holds for the bcc and fcc lattices.

Problem 8-7. Find the volume of the first Brillouin zone in sodium (Na) and in copper (Cu). Answer: 6.24 Å^{-1}, 21.1 Å^{-1}.

The number of states in the first Brillouin zone can now be calculated:

$$V_k\, g(\mathbf{k}) = \left[(2\pi)^3/V_c\right]\left[2V/(2\pi)^3\right] = 2V/V_c. \qquad (8\text{-}15)$$

The ratio V/V_c is the total volume of the crystal divided by the volume of each primitive unit cell. This gives us simply the number N_c of primitive unit cells in the crystal. Therefore, the number of states in the first Brillouin zone is equal to $2N_c$, twice the number of primitive unit cells in the crystal. This is the number of states in a single energy band.

States are filled with conduction electrons just like in the free-electron model. Using Pauli's exclusion principle, we fill the states starting with those of lowest energy until we have put in all the electrons. For example, sodium metal (Na) has a bcc lattice with one atom in each primitive unit cell. Since each atom contributes one conduction electron, the number of these electrons must be equal to N_c, the number of primitive unit cells (which is the number of atoms in this case). Since each energy band contains $2N_c$ states, there are enough conduction electrons to fill the first energy band *half way*. The density of occupied states would look something like Fig. 8-8. The Fermi energy E_F, as in the free-electron model, is the energy of the highest occupied states. The Fermi energy is shown as a dashed horizontal line in Fig. 8-6.

As another example, consider barium metal (Ba). It has a bcc lattice with one atom in each primitive unit cell. Each Ba atom contributes *two* conduction electrons. Thus, the number of these electrons is equal to $2N_c$, which is exactly the number of available states in each energy band. There are enough electrons to exactly fill the first band. In Fig. 8-9 we show the first two bands in Ba. We see that the energy at the top of the first band in the [111] direction is about 3.0 eV, while the energy at the bottom of the second band in the [100] direction

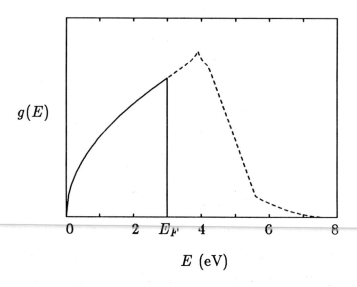

Fig. 8-8. The density of occupied states in sodium metal. The dashed line is the density of unoccupied states in the first Brillouin zone.

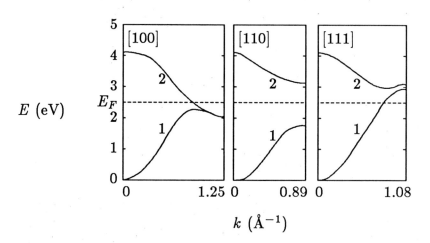

Fig. 8-9. The first two energy bands in barium metal. The first Brillouin zone is the same as that of sodium, shown in Fig. 8-6. See, for example, G. Johansen, *Solid State Commun.* **7**, 731 (1969).

is about 2.0 eV. The first and second bands *overlap*. Some of the states in the second band have a lower energy than some of the states in the first band. Thus, when we fill the states with electrons, some of them must be put into the second band and consequently there are not enough of them to completely fill the first band. The Fermi energy in Ba is equal to about 2.5 eV, as shown in the figure. We can see that some of the states near the bottom of the second band are occupied, and some of the states near the top of the first band are unoccupied.

Problem 8-8. Consider a piece of sodium metal (Na) of mass 1.00 g. (a) How many electron states are in each band? (b) How many conduction electrons are there? Answer: 5.24×10^{22}, 2.62×10^{22}.

Problem 8-9. The compound CuZn is a metal with the CsCl structure. Each Cu atom contributes one conduction electron and each Zn atom contributes two conduction electrons. How many energy bands can be filled by the conduction electrons in CuZn? Answer: $1\frac{1}{2}$ bands.

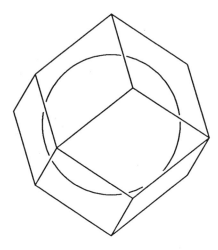

Fig. 8-10. Fermi surface in sodium metal.

8-5 Fermi Surface

In sodium metal, the occupied states follow the free-electron model quite closely, and the Fermi surface is very nearly a sphere, as in the free-electron model. The sphere sits inside the first Brillouin zone, as shown in Fig. 8-10. All states inside the sphere are occupied, and all states outside the sphere are unoccupied. Copper metal (Cu) is a good example of how the Fermi surface can be distorted from the sphere shape. In Fig. 8-11, we see the energy band in Cu which the conduction electrons fill. We see that it is highly distorted from the free-electron model. In fact, the Fermi level (the horizontal dashed line) lies *above* the top of the band in the [111] direction. This means that the Fermi surface touches the first Brillouin zone boundary in the [111] direction. The Fermi surface is shown in Fig. 8-12. It is somewhat spherical but with protrusions out to the zone boundary in the eight [111] directions. The Fermi surfaces of most metals have even more complicated shapes.

8-6 Atomic Model

As a last topic in this chapter, we will show how energy bands arise from an "atomic model" of solids. The simplest atom is hydrogen. It has one proton as a nucleus and one electron. If we consider the proton to be stationary and solve Schroedinger's equation for the electron, we find that the electron is only permitted to have certain energies (see Fig. 8-13). The electron normally occupies the lowest energy state ($E = -13.6$ eV).

We get a similar result for other types of atoms which have more than one electron. The electrons are only permitted to have certain energies. Similar to solids, we can consider the atom to contain electron *states* which we then fill up with electrons, starting with the lowest energy. From Pauli's exclusion principle, only one electron is allowed to occupy a state at a time, and thus for atoms with many electrons, we must put some electrons into states of quite high energy because all the states of lower energy are already filled. States (or **orbitals**, as they are usually called) in atoms are labeled with symbols

CHAPTER 8 BAND THEORY OF METALS   167

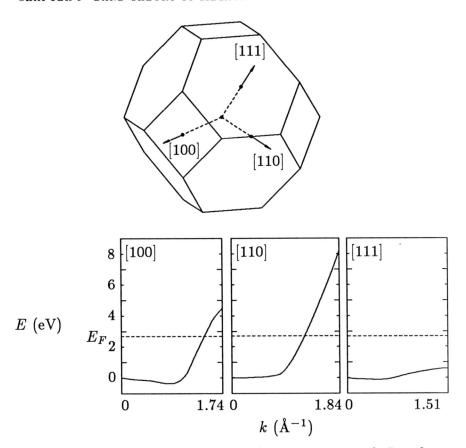

Fig. 8-11. Conduction-electron band in copper metal. See, for example, G. A. Burdick, *Phys. Rev.* **129**, 138 (1963).

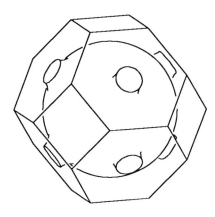

Fig. 8-12. The Fermi surface of copper metal.

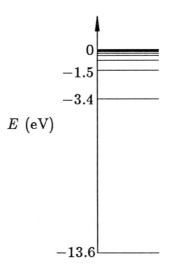

Fig. 8-13. The energy levels of an electron in the hydrogen atom.

Table 8-1. Labeling of atomic orbitals and the number of orbitals of each type. They are given in the usual order of their energy, with the lowest energy first.

1s	2
2s	2
2p	6
3s	2
3p	6
4s	2
3d	10
⋮	⋮

such as 1s, 2s, 2p, 3d, etc. There is also more than one orbital of each type. For example, there are two 1s orbitals and six 2p orbitals. In Table 8-1 are shown the labels of atomic orbitals, in the usual order of their energy.

For example, a lithium atom (Li) has three electrons. Two of them occupy 1s orbitals and one occupies a 2s orbital. Suppose that we bring two Li atoms close together so that they can interact. If we solve Schroedinger's equation, we find that there are now *four* 1s orbitals, *four* 2s orbitals, *twelve* 2p orbitals, etc. Since there are six electrons, four of them occupy the 1s orbitals, and two occupy the 2s orbitals. We also find that the energies of the different orbitals are slightly split apart so that the energy levels look something like Fig. 8-14a.

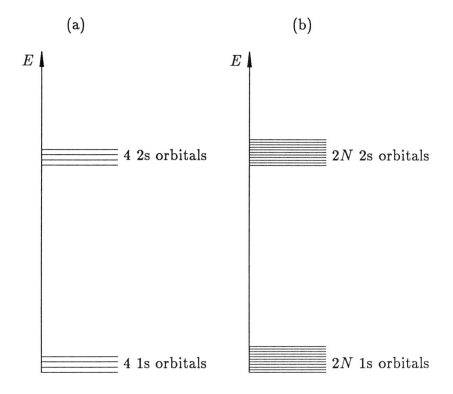

Fig. 8-14. (a) Orbitals for two Li atoms brought close together. (b) Orbitals for N Li atoms brought close together.

If we bring N lithium atoms close together, we obtain $2N$ 1s orbitals, $2N$ 2s orbitals, $6N$ 2p orbitals, etc. The energies of these orbitals are also all slightly split apart from each other. However, if N is very large, like in a crystal, these levels are very close to each other and form "bands" of energy (see Fig. 8-14b). Since there are $3N$ electrons, they fill the 1s band completely and fill the 2s band half way. The electrons in the 2s band are the conduction electrons (one per Li atom). This example illustrates how energy bands can be understood in terms of atomic energy levels. More complicated cases can also be explained this way, but to pursue this further would be beyond the scope of this book.

CHAPTER 9

ELECTRICAL CONDUCTIVITY OF METALS

9-1 Group Velocity

In this chapter, we will discuss electrical conductivity of metals using the band theory developed in the previous chapter. As usual, we will first treat it in one dimension. In Chapter 6, we found that the velocity v of a free electron is equal to its group velocity $v_g = d\omega/dk$ [Eq. (6-22)]. This relation also holds for electrons which are not free, such as those occupying states in an energy band. Using $E = \hbar\omega$, we can write the electron's velocity as

$$v = \frac{d\omega}{dk} = \frac{1}{\hbar}\frac{dE}{dk}. \qquad (9\text{-}1)$$

Since the energy E of an electron in a solid does not depend on k in the same way as for a free electron, we do *not* simply get $v = \hbar k/m$. Thus, $\hbar k$ is not the true momentum of the electron.

Consider an energy band such as that shown in Fig. 9-1. From Eq. (9-1), we see that the velocity of the electron is proportional to the slope of this curve. Using Eq. (9-1), we plot v in Fig. 9-2. Note that the velocity of the electron is zero at the top of the band as well as at the bottom. As we discussed in Chapter 8, this behavior is due to Bragg reflection at the first Brillouin zone boundary. The electron cannot travel through the metal with that de Broglie wavelength (see Problem 8-1).

Problem 9-1. Consider an electron in the first energy band of barium metal (Ba) with a wave vector **k** along the [100] direction. From Fig. 8-9, find the approximate values of k where the velocity of the electron is zero.

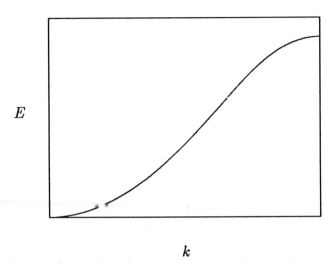

Fig. 9-1. An energy band in a one-dimensional metal.

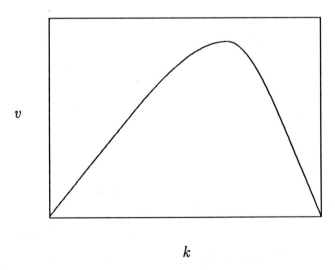

Fig. 9-2. Velocity of electrons in the band shown in Fig. 9-1.

9-2 Equation of Motion

Consider next the effect of some external force F_{ext} such as an applied electric field or magnetic field. This force does work on the electron. If the electron travels a small distance dx, the work done is $F_{ext}dx$. This changes the energy of the electron by an amount,

$$dE = F_{ext}dx. \tag{9-2}$$

If the electron has velocity v, then the distance traveled in time dt is $dx = vdt$. We thus obtain

$$dE = F_{ext}vdt, \tag{9-3}$$

or

$$\frac{dE}{dt} = F_{ext}v, \tag{9-4}$$

which is the rate at which the energy of the electron changes due to the external force.

Using Eq. (9-1), we then obtain

$$\frac{dE}{dt} = F_{ext}\frac{1}{\hbar}\frac{dE}{dk}. \tag{9-5}$$

Solving for F_{ext}, we obtain

$$F_{ext} = \hbar\frac{dk}{dt}. \tag{9-6}$$

For a free electron, where $p = \hbar k$, this is just Newton's law,

$$F = \frac{d(\hbar k)}{dt} = \frac{dp}{dt} = m\frac{dv}{dt} = ma. \tag{9-7}$$

However, for an electron in a metal, $\hbar k$ is not the true momentum of the electron, and, furthermore, F_{ext} is not the *total* force on the electron. There are internal forces acting on the electron as well, such as the interaction with the positive ions. Nevertheless, Eq. (9-6) is very useful. We usually do not know the electron's true momentum, and we usually do not know the internal forces on the electron. However, we usually *do* know the external forces, and we know the wave vector **k** of the electron in a given state. In three dimensions, we obtain an expression very much like Eq. (9-6):

$$\mathbf{F}_{\text{ext}} = \hbar \frac{d\mathbf{k}}{dt}. \qquad (9\text{-}8)$$

9-3 Electrical Conductivity

Electrical conductivity is easily explained in terms of Eq. (9-8). If we apply an electric field $\mathcal{E}$, the external force on an electron is equal to $-e\mathcal{E}$, and we thus obtain from Eq. (9-8)

$$\frac{d\mathbf{k}}{dt} = -(e/\hbar)\mathcal{E}. \qquad (9\text{-}9)$$

This equation states that an electric field $\mathcal{E}$ causes the wave vector **k** of an electron to change. This means that the electron moves to new states. Imagine that in **k**-space each electron occupies some "position" equal to the wave vector **k** of the state it occupies. Then, the quantity $d\mathbf{k}/dt$ is the "velocity" of the electron in **k**-space. We can interpret Eq. (9-9) to mean that in a constant electric field $\mathcal{E}$, the velocity of each electron in **k**-space is equal to a constant. All the electrons move with uniform velocity through **k**-space. As a consequence, the entire Fermi surface moves through **k**-space, keeping its original shape.

The motion of electrons through **k**-space is also influenced by another effect, though. Electrons are constantly colliding with "obstacles" in the crystal. On the average, these collisions try to bring the electrons back to states of minimum energy, moving the Fermi surface back toward its original undisplaced position. The collisions thus oppose the action of the electric field which tries to move the Fermi surface *away* from its original undisplaced position. At equilibrium, both effects cancel each other, and the Fermi surface sits at rest in **k**-space with some net displacement $\Delta\mathbf{k}$.

In the free-electron model, a displacement $\Delta \mathbf{k}$ of the Fermi surface gives rise to uncompensated electrons and consequently a net current (see Figs. 7-8 and 7-9). In the band model, we get a similar result, except that the Fermi surface in general is no longer a sphere. Consider, for example, the Fermi surface of copper (Cu) shown in Fig. 8-12. A cross section of this Fermi surface is shown in Fig. 9-3. If we apply an electric field $\mathcal{E}$ pointing to the left, the Fermi surface will be displaced to the right, as shown in Fig. 9-4.

Since some of the states at the boundary of the first Brillouin zone in Cu are occupied, a displacement of the Fermi surface in this case causes some of the electrons to go outside the first Brillouin zone. We can relabel the wave vectors $\mathbf{k}$ of these electrons so that they are *inside* the first Brillouin zone, though, by adding some reciprocal lattice vector $\mathbf{G}$. For example, in Fig. 9-4 we see a point outside the first Brillouin zone brought back inside using a reciprocal lattice vector $\mathbf{G}$. Actually, these electrons did not really *leave* the first Brillouin zone, but simply jumped to the opposite side when they crossed the boundary. (Furthermore, these electrons remain in the same energy band. Energy gaps at the first Brillouin zone keep electrons from moving into a new band when they cross the zone boundary.) If we redraw Fig. 9-4, relabeling all electron states so that they are inside the first Brillouin zone, we obtain a Fermi surface which looks like Fig. 9-5.

With the Fermi surface displaced, some of the electrons do not come in $+\mathbf{k}$ and $-\mathbf{k}$ pairs and are thus "uncompensated." The uncompensated electrons in Cu are shown by the shaded area in Fig. 9-6. Note that in the case of Cu, some of the uncompensated electrons are on the left side as well as the right side. However, most of them are on the right side, and the average wave vector $\mathbf{k}$ of the electrons points to the right, giving rise to an electric current to the left, as in the case of the free-electron model.

The size of the electric current depends on the number of uncompensated electrons, which, in turn, depends on the displacement of the Fermi surface as well as the shape and

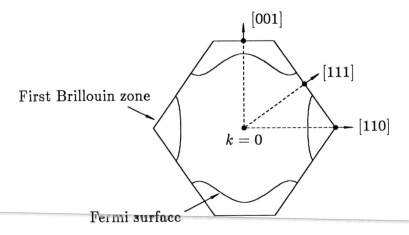

Fig. 9-3. Cross section of the Fermi surface of copper.

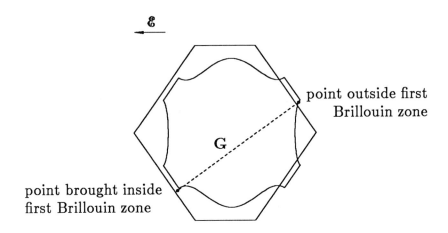

Fig. 9-4. Displacement of the Fermi surface of copper in an electric field $\mathcal{E}$. A point shown outside the first Brillouin zone is brought back inside using a reciprocal lattice vector **G**.

CHAPTER 9   ELECTRICAL CONDUCTIVITY OF METALS   177

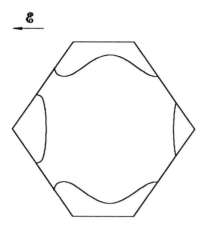

Fig. 9-5. Displaced Fermi surface in the first Brillouin zone of copper.

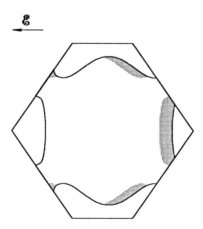

Fig. 9-6. Uncompensated electrons in copper.

area of the Fermi surface. This explains why the electrical conductivity σ of aluminum (Al) is about the same size as in Cu, even though Al has *three* times as many conduction electrons as Cu. The area of the Fermi surface in Al is about the same as in Cu. Thus, they have about the same number of uncompensated electrons carrying the current.

Problem 9-2. Consider an electron at the Fermi level in copper (Cu) with a wave vector **k** in the [100] direction (see Fig. 8-11). We apply an electric field $\mathscr{E} = 1.00$ V/m in a negative [100] direction. If the electron could travel without hindrance (no collisions), how long would it take to reach the boundary of the first Brillouin zone? Answer: 2 μs.

9-4 Metals, Insulators, Semiconductors

We can use the band model of metals to explain why some crystals are insulators and do not conduct an electric current. Consider an energy band which is completely filled with electrons. This means that *every* electron state in the first Brillouin zone is occupied. If we apply an electric field $\mathscr{E}$, the electrons move through **k**-space, as given by Eq. (9-9), but those that move across the first Brillouin zone boundary actually jump to the other side of the zone into states left unoccupied by electrons that are moving *away* from the boundary. All the states in the first Brillouin zone are still occupied. There will never be any uncompensated electrons since states in the first Brillouin zone always come in $+\mathbf{k}$ and $-\mathbf{k}$ pairs. A completely filled band *cannot* carry a current.

Thus, in order for a given crystal to be able to carry a current, at least one of its energy bands must be only partially filled. This is why sodium (Na) and copper (Cu) are metals, for example. They both have an energy band which is only half-filled with electrons. Remember that the number of states in a band is equal to $2N_c$, twice the number of unit cells in the crystal. If a crystal has an *odd* number of electrons in each unit cell, then at least one energy band will be only partially filled,

CHAPTER 9 ELECTRICAL CONDUCTIVITY OF METALS   179

and the crystal must be a metal. In Na, there are 11 electrons in every unit cell. In Cu, there are 29. These are odd numbers. Thus, Na and Cu must be metals.

On the other hand, in sodium chloride (NaCl) each unit cell contains one Na atom and one Cl atom and thus contains $11 + 17 = 28$ electrons. This is an even number. There are just the right amount of electrons in NaCl to completely fill 14 bands (including the bands formed by the core electrons). All of the bands in NaCl are either completely filled or completely empty. NaCl cannot carry a current. It is an **insulator**.

In barium (Ba), each unit cell contains 56 electrons, which is an *even* number. But we know that Ba is a metal. As we saw in Chapter 8, the overlapping of energy bands prevents the electrons from completely filling the last band (see Fig. 8-9). Instead, two different bands are partially filled. Generally, if there is an *odd* number of electrons in the unit cell, the crystal *must* be a metal. However, if there is an *even* number of electrons in the unit cell, the crystal may be either a metal or an insulator, depending on whether the last band overlaps with the next higher band. In Ba, the bands overlap, but in NaCl, they do not. In NaCl, the last filled band is separated from the next higher band by an energy gap. This is why NaCl is an insulator.

We must be careful how we apply this rule. For example, the hydrogen atom has one electron. Hydrogen atoms form a crystalline solid at temperatures below 14 K. Solid hydrogen is *not* a metal, even though there is an odd number of electrons per *atom*. Solid hydrogen has a structure with *two* atoms per unit cell and thus *two* electrons per unit cell.

As one last example, consider some crystals with the diamond structure: carbon (C), silicon (Si), and germanium (Ge). In C, each unit cell contains two atoms and thus contains 12 electrons. Its last filled band is separated from the next higher band by 5.3 eV. C is an insulator. In Si and Ge, each unit cell also contains an even number of electrons (28 in Si and 64 in Ge), but the last filled band is separated from the next higher band by a much smaller amount (1.1 eV in Si and

0.7 eV in Ge). At room temperature, some electrons have enough thermal energy to jump across the gap and occupy states in the empty band. These electrons can carry a current. Such crystals are called **semiconductors**. We will discuss their properties in great detail in the following chapters.

Problem 9-3. In Appendix 3 are listed a number of elements with a bcc lattice. Which of these *must* be metals because of the number of electrons in each unit cell?

Problem 9-4. In Appendix 3 are listed a number of compounds with a CsCl structure. Which of these *must* be metals because of the number of electrons in each unit cell?

9-5 Electron Collisions

Earlier in this chapter, we discussed how collisions of electrons limit the electrical conductivity of a metal. What kind of obstacles do the electrons collide with? To answer this, consider a crystal which is *perfectly* periodic. The contents of every unit cell are identical at every instant of time. Because of the wave properties of electrons, when they travel through such a crystal, they can only be scattered if Bragg's law is satisfied. This only occurs for electrons in states at the first Brillouin zone boundary. Thus, electrons in states inside the first Brillouin zone should be able to travel through such a crystal without any hindrance at all.

Real crystals are not perfectly periodic, however. There are two main types of imperfections in crystals. First, real crystals always contain some impurities (for example, there may be some Fe atoms in a Cu crystal). Second, lattice vibrations are always present. At any given instant of time, ions are slightly displaced in a random way from their positions in a "perfect" crystal. Electrons can collide with these imperfections in the crystal. The greater the degree of imperfection, the greater the probability will be that an electron will collide with one of them. This results in a smaller electrical conductivity.

Problem 9-5. Explain why liquid metals always have a lower conductivity than solid metals of the same type.

Problem 9-6. Explain why the conductivity of an alloy is generally lower than that of its constituent metals.

The effect of impurities on conductivity is generally temperature independent. The effect of lattice vibrations, however, is strongly temperature dependent. The ions vibrate with larger amplitude at high temperatures and thus scatter the conduction electrons more often. In general, the resistivity $\rho = 1/\sigma$ of a metal due to lattice vibrations is linear in the temperature T. The total resistivity is given by the sum of the resistivities due to the impurities and lattice vibrations, and is shown in Fig. 9-7 for a particular sample of sodium metal (Na). We see that at room temperature the effect of lattice vibrations dominates, and ρ is linear in T. At low T, the impurities dominate, and ρ is T-independent. The value of ρ at low T depends on the impurity content of the metal and varies from sample to sample. The value of ρ at high T, however, depends only on the amplitude of the lattice vibrations and is identical for all samples of Na.

In Problem 7-12, we saw that conduction electrons near the Fermi surface in sodium metal travel, on the average, $\ell = 330$ Å between collisions. We can now understand this result, in view of our present model. Electrons do not collide with just any Na ion, but only with those Na ions which are displaced far enough away from their position of equilibrium. At lower temperatures, the distance between collisions is even greater. We can now see that the picture given in Fig. 4-1 of an electron colliding with ions in a perfectly periodic array is false.

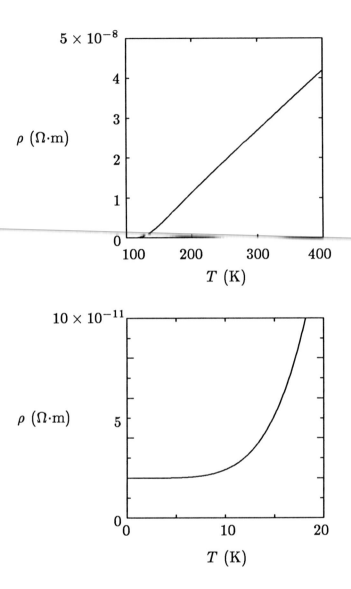

Fig. 9-7. The resistivity of a particular sample of sodium metal.

Problem 9-7. Find the average distance ℓ between collisions of conduction electrons in the sample of sodium metal in Fig. 9-7 at $T = 4$ K. Answer: 70 µm.

9-6 Effective Mass

If we apply an external force F_{ext} on an electron in a metal, what is the electron's *true* acceleration? From calculus, we have

$$a = \frac{dv}{dt} = \frac{dv}{dk}\frac{dk}{dt}. \tag{9-10}$$

Using Eqs. (9-1) and (9-6), we obtain

$$a = \left(\frac{1}{\hbar}\frac{d^2 E}{dk^2}\right)\left(\frac{1}{\hbar}F_{\text{ext}}\right). \tag{9-11}$$

This would look like Newton's law, $a = F/m$, if we defined an **effective mass** m^* of the electron,

$$m^* = \left(\frac{1}{\hbar^2}\frac{d^2 E}{dk^2}\right)^{-1}. \tag{9-12}$$

Then we have

$$a = F_{\text{ext}}/m^*. \tag{9-13}$$

Thus, we can find the acceleration of the electron without needing to know explicitly what the internal forces are. We only need to know what the effective mass m^* is. The electron in the solid responds to the external force as though it were a *free* electron of mass m^*.

We can calculate the effective mass m^* of an electron in a given state **k** using Eq. (9-12) if we know the band structure. For the energy band shown in Fig. 9-1, we calculate m^* using Eq. (9-12) and plot the result in Fig. 9-8. At the top of the band, we see that m^* is negative! An external force trying to push the electron toward the top of the band causes the

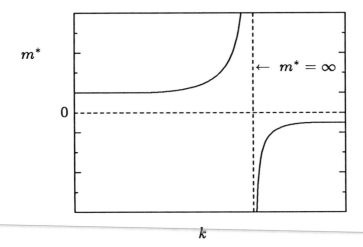

Fig. 9-8. Effective mass of the electrons in the energy band of Fig. 9-1.

velocity of the electron to *decrease* (see Fig. 9-2). Thus, a force in a positive k direction causes a *deacceleration* (negative acceleration). The electron behaves as though it had a negative mass. When we "push" on it, it slows down instead of speeding up. And the harder we push, the faster is slows down. This behavior is due entirely to internal forces which more than cancel the external force which we apply.

We also see in Fig. 9-8 a point near the middle of the band where the effective mass is infinite! This happens where the velocity of the electron is at a maximum in Fig. 9-2. Since the slope of v is zero there, a force applied to the electron causes k to change but not v. If a force on an object cannot change its velocity, the object behaves as though it had infinite mass.

Generally, if we are given a plot of E as a function of k, we can tell by inspection where the effective mass is positive, negative, or infinite. From calculus, we know that the sign of the second derivative of a function tells us whether the function is concave up or down. From Eq. (9-12) we see that m^* is positive where $E(k)$ is concave up, m^* is negative where $E(k)$

is concave down, and m^* is infinite at points where the curve changes from concave up to concave down or vice versa (called "points of inflection" in calculus).

For example, from the plot of $E(k)$ for Ba along the [100] direction (see Fig. 8-9), we can see that an electron in the first energy band has positive effective mass for values of k between 0 and about 0.5 Å^{-1} and has negative effective mass for values of k between 0.5 Å^{-1} and the zone boundary at 1.25 Å^{-1}. An electron at $k \cong 0.5$ Å^{-1} has an infinite effective mass.

Problem 9-8. Consider an electron in the *second* energy band of Ba with a wave vector **k** along the [100] direction (see Fig. 8-9). For what range of values of k will the effective mass of the electron be positive? Negative? Infinite?

9-7 Holes

There is an alternate way to describe electric currents arising from electrons in a given energy band. Let us remove all the electrons from the occupied states and put positively charged particles into the formerly unoccupied states. These positively charged particles are called **holes**. In Fig. 9-9, we show diagrammatically what we mean. The circle represents the Fermi surface. The square represents the boundary of the first Brillouin zone. In case (a), the states inside the Fermi surface are occupied by electrons. The states outside the Fermi surface are unoccupied. In case (b), the states inside the Fermi surface are *unoccupied*. The states outside the Fermi surface are occupied by *holes*. The velocity of a hole in a given state is equal to the velocity an electron would have if it occupied that state. The charge of a hole is equal in magnitude to that of an electron, but positive instead of negative.

If we apply an electric field $\mathscr{E}$ pointing to the left, the Fermi surface is displaced to the right, as shown in Fig. 9-10. This happens because the electrons move to the right, in agreement with Eq. (9-9). From Fig. 9-10 we see that the holes also move

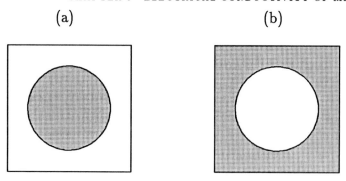

Fig. 9-9. Occupied states in an energy band (the square) represented in two different ways. (a) Electrons fill states inside the Fermi surface (the circle), and (b) holes fill states outside the Fermi surface.

to the right. However, since the charge of a hole is positive, the force must be the *same* direction as $\mathcal{E}$. Thus, we must have for holes,

$$\mathbf{F}_{\text{ext}} = -\hbar \frac{d\mathbf{k}}{dt}, \qquad (9\text{-}14)$$

which has the opposite sign as Eq. (9-8) for electrons.

The displacement of the Fermi surface causes a current

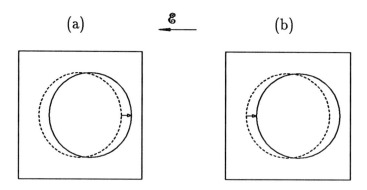

Fig. 9-10. Displacement of the Fermi surface by an electric field $\mathcal{E}$. Electrons and holes are *both* displaced in a direction opposite to $\mathcal{E}$.

to flow (see Fig. 9-11). In case (a), the current is caused by uncompensated electrons on the right-hand side of the Fermi surface. These uncompensated electrons have a velocity to the right, and thus we get a net flow of electrons to the right. In case (b), the current is caused by uncompensated holes on the *left*-hand side of the Fermi surface. These uncompensated holes have a velocity to the *left*, and thus we get a net flow of holes to the left. Here we can see that the two models of electrical conductivity are equivalent. A flow of positively charged particles to the left is equivalent to a flow of negatively charged particles to the right. In both cases, we get an electric *current I* to the left, the same direction as $\mathscr{E}$.

We can determine the acceleration of a hole due to an external force, just as we did for electrons in Eq. (9-13). However, we must use $F_{\text{ext}} = -\hbar dk/dt$ for holes instead of Eq. (9-6) for electrons. If we do this, we get

$$\mathbf{a} = -\mathbf{F}_{\text{ext}}/m^*, \qquad (9\text{-}15)$$

where m^* is the effective mass an electron would have if it occupied that state. We can make this equation look like Newton's law again if we define the effective mass m_p^* of a hole to

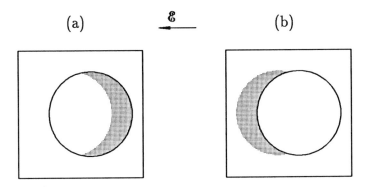

Fig. 9-11. Electric current caused by (a) uncompensated electrons moving to the right or (b) uncompensated holes moving to the left.

be
$$m_p^* = -m_n^* = -\left(\frac{1}{\hbar^2}\frac{d^2E}{dk^2}\right)^{-1}. \qquad (9\text{-}16)$$

Here, we write the effective mass of the electron as m_n^* (n for "negative") to distinguish it from the effective mass m_p^* of the hole (p for "positive").

Problem 9-9. Show that the acceleration of a hole in a one-dimensional metal is given by Eq. (9-15). (Start with $F_{\text{ext}} = -\hbar dk/dt$ and $v = d\omega/dk$.)

The effective mass of a hole in a given state is equal in magnitude but *opposite in sign* to the effective mass an electron would have if it occupied that state. For example, let us suppose that the electrons near the right-hand side of the Fermi surface in Fig. 9-10(a) have a positive effective mass. The electric field exerts a force to the right. This moves the electrons to the right, into states where they have greater velocity to the right. Thus, the acceleration is in the same direction as the force, as expected. On the other hand, holes near the same Fermi surface [see Fig. 9-10(b)] experience a force to the *left*. But they nevertheless move to the *right*, into states where they have greater velocity to the *right*. The acceleration is thus to the right while the force is to the left. They behave like particles of *negative* mass.

Problem 9-10. Using the same line of reasoning as in the above paragraph, explain why the effective mass of a hole in a given state is positive if an electron in the same state would have a negative effective mass.

When the electric current in a band arises from electrons with negative effective mass, it is very convenient to use the "hole model" we have described above. Then, instead of negatively charged particles with negative mass, we have positively charged particles with positive mass. This turns out to be a much better physical description of the electric current.

9-8 Hall Effect

Consider, for example, a metal which has an energy band nearly full of electrons. The Fermi level is near the top of the band where the effective mass of the electrons is negative. The electric current arises from the flow of uncompensated electrons with negative effective mass near the Fermi surface. In the "hole model," the electric current arises from the flow of uncompensated *holes* with *positive* effective mass near the Fermi surface. These holes can actually be detected, using the Hall effect. We discussed the Hall effect using the classical model in Chapter 4. A current of electrons placed in a magnetic field **B** gives rise to an electric field, called the Hall field $\mathcal{E}_H$, given by Eq. (4-17),

$$\mathcal{E}_H = -R_H \mathbf{J} \times \mathbf{B}. \tag{9-17}$$

In the classical model, the Hall coefficient R_H is given by Eq. (4-18),

$$R_H = -1/ne. \tag{9-18}$$

This was derived for a current of *electrons*, each with charge $-e$. If we consider a current of *holes* with charge $+e$ and density p, we obtain from the classical model

$$R_H = +1/pe. \tag{9-19}$$

The Hall field $\mathcal{E}_H$ given by Eq. (9-17) should thus be in the *opposite* direction for a current of holes. This is exactly what happens. In Appendix 4, we find a number of metals, such as Fe and Zn, which have a positive Hall coefficient R_H. In these metals, the electric current is dominated by electrons with negative effective mass. We say that the current consists of positively charged holes in these metals.

Problem 9-11. Show that the Hall coefficient R_H for a current of positively charged particles is given by Eq. (9-19) in the classical model.

The expressions in Eqs. (9-18) and (9-19) are very inadequate for calculating the Hall coefficient in real metals except in cases where the Fermi surface is spherical, as in sodium metal (see Problem 4-6).

9-9 Cyclotron Resonance

The effective mass of electrons is measured experimentally with cyclotron resonance. In Chapter 4, we found that free electrons in the presence of a magnetic field B move in circles with frequency $\omega = eB/m$ [Eq. (4-21)]. This motion can be detected by absorption of electromagnetic radiation of the same frequency. In a real solid, the electrons in an energy band behave just like free electrons, but with an effective mass m^*. Thus, in a magnetic field B, they will move in circles with frequency,

$$\omega = eB/m^*. \tag{9-20}$$

Problem 9-12. A cyclotron resonance is observed in lead metal (Pb) at 8900 MHz for a field $B=0.24$ Tesla. Find the effective mass of the detected electron in terms of the mass m of a free electron. Answer: 0.75m.

Note that with cyclotron resonance, we can only measure the magnitude of the effective mass, not its sign.

CHAPTER 10

SEMICONDUCTORS

10-1 Band Structure

Semiconductors have vast technological importance. The study of the behavior of electrons in semiconductor crystals has led to the design of electronic devices which have revolutionized our way of life.

In a semiconductor crystal, there are an *even* number of electrons in each primitive unit cell. Also, there is an energy gap, of magnitude E_g, between the highest occupied energy band and the next higher band. Consequently, at $T = 0$, all bands are either completely filled or completely empty. There are no partially filled bands. The highest occupied band is called the **valence band** (VB), and the next higher band is called the **conduction band** (CB). At $T = 0$, the CB is completely empty. The gap energy E_g is the amount of energy required for an electron to go from the top of the VB to the bottom of the CB.

At room temperature, if the energy gap E_g is not too large, some electrons near the top of the VB may have enough thermal energy to jump across the gap and occupy states near the bottom of the CB. The electrons that jump into the CB leave behind unoccupied states at the top of the VB. Since the electrons near the top of a band generally have negative effective mass, we will use the "hole model" for the VB. We will consider all states in the VB to be "unoccupied" except for a few near the top of the band which are occupied by holes with *positive* effective mass. The electrons in the CB as well as the holes in the VB can carry an electric current. A number of different semiconductors and their gap energies are listed in Appendix 5.

10-2 Simple Model

Let us use the following simplified model of the VB and CB bands in a semiconductor. Let the maximum energy of the

VB and the minimum energy of the CB both occur at $k = 0$ (similar to the second and third bands shown in Fig. 8-4 for the one-dimensional metal). Since we are always free to choose where $E = 0$, let us choose $E = 0$ at the top of the VB. Then at the bottom of the CB, we have $E = E_g$.

Since the electrons only occupy states near the bottom of the CB where $k \cong 0$, we can expand E in powers of k:

$$E(k) = E(0) + \left.\frac{dE}{dk}\right|_{k=0} k + \frac{1}{2} \left.\frac{d^2 E}{dk^2}\right|_{k=0} k^2 + \cdots \qquad (10\text{-}1)$$

Since the value of k is small, we will only keep terms up to k^2. $E(0)$ is the energy at $k = 0$ which is simply the energy at the bottom of the CB, i.e., $E(0) = E_g$. Since $E(k)$ has a minimum at $k = 0$, the first derivative dE/dk must be zero there. The second derivative can be written in terms of the effective mass m_n^* of the electron at the bottom of the CB. From Eq. (9-12), we obtain $d^2 E/dk^2 = \hbar^2/m_n^*$. Thus, Eq. (10-1) becomes, for the CB,

$$E = E_g + \frac{\hbar^2}{2m_n^*} k^2. \qquad (10\text{-}2)$$

For the VB, we can also expand E in powers of k as in Eq. (10-1). In this case, though, $E(0) = 0$, and E has a maximum at $k = 0$. Also, from Eq. (9-16), we obtain $d^2 E/dk^2 = -\hbar^2/m_p^*$, where m_p^* is the effective mass of the hole at the top of the VB [not necessarily equal to the m_n^* in Eq. (10-2)]. Thus, Eq. (10-1) becomes, for the VB,

$$E = -\frac{\hbar^2}{2m_p^*} k^2. \qquad (10\text{-}3)$$

Using Eqs. (10-2) and (10-3), we plot in Fig. 10-1 the band structure near the energy gap of the model semiconductor. The electrons near the bottom of the CB and the holes near the top of the VB behave very much like free particles.

CHAPTER 10 SEMICONDUCTORS

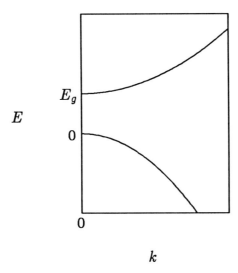

Fig. 10-1. Band structure of a model semiconductor near the energy gap.

Problem 10-1. Using Eq. (9-1), show that the velocity of the electrons near the bottom of the CB is $v = \hbar k/m_n^*$ and the velocity of the holes near the top of the VB is $v = -\hbar k/m_p^*$. How does this result show us that these electrons and holes behave like free particles?

We can use the results of the free-electron model almost directly here. From Eq. (7-16) we obtain the density of states for the CB:

$$g(E) = \frac{V}{2\pi^2}(2m_n^*/\hbar^2)^{3/2}(E - E_g)^{1/2}. \qquad (10\text{-}4)$$

Notice that we substituted $E - E_g$ for E in Eq. (7-16) since the meaning of E in that equation is the energy *above* the state at $k = 0$. We chose the energy at $k = 0$ to be equal to E_g. For the VB, we obtain a similar result. Here we can use Eq. (7-16) if we interpret E in that equation to mean the energy *below*

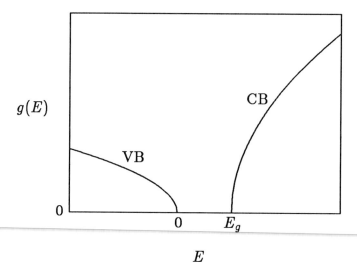

Fig. 10-2. The density of states for the model semiconductor.

the state at $k = 0$. We thus obtain the density of states for the VB:

$$g(E) = \frac{V}{2\pi^2}(2m_p^*/\hbar^2)^{3/2}(-E)^{1/2}. \qquad (10\text{-}5)$$

In Fig. 10-2, we plot Eqs. (10-4) and (10-5) for the model semiconductor. This figure again illustrates the free-particle nature of these electrons and holes. Each curve is a parabola on its side, just like $g(E)$ for a free particle shown in Fig. 7-3.

10-3 Density of Electrons and Holes

At $T = 0$, all the states in the VB are occupied, and all the states in the CB are unoccupied. For $T > 0$, some electrons have enough thermal energy to go into the CB, leaving holes behind in the VB. Since every hole in the VB was left behind by an electron that jumped into the CB, the total number of holes in the VB must be equal to the total number of electrons in the CB. (This is only strictly true for a *pure* semiconductor, which we are considering here.)

The density of *occupied* states is given by $f_D(E)g(E)$, where $f_D(E)$ is the Fermi-Dirac distribution function in

Eq. (7-27):
$$f_D(E) = \frac{1}{\exp\left(\frac{E-E_F}{k_B T}\right) + 1}. \qquad (10\text{-}6)$$

This function gives us the probability that a state at energy E will be occupied. Let us consider a Fermi energy E_F near the center of the gap (Fig. 10-3). The resulting density of states occupied by electrons is plotted in Fig. 10-4. The probability that a state will be *unoccupied* is equal to $1 - f_D(E)$, which we plot in Fig. 10-5. The product $[1 - f_D(E)]g(E)$ is thus the density of *unoccupied* states or, for the VB, the density of states occupied by *holes*. We plot this in Fig. 10-6.

By definition of density of states, the *area* under the curves in Figs. 10-4 and 10-6 gives us the number of states. In a pure semiconductor, the number of holes in the VB must be equal to number of electrons in the CB. This means that the area under the "VB" curve in Fig. 10-6 must equal the area under the "CB" curve in Fig. 10-4. The Fermi energy E_F was chosen to make these two areas equal. If E_F were larger, then there would be more electrons in the CB than holes in the VB. If E_F were smaller, then there would be more holes in the VB than electrons in the CB. As we can see, the Fermi energy must be approximately in the center of the energy gap. The requirement that the number of holes in the VB equal the number of electrons in the CB determines the Fermi energy E_F.

Let us now calculate the number of electrons and holes. The number of electrons in the CB is given by the area under the "CB" curve in Fig. 10-4:

$$N = \int_{E_g}^{\infty} f_D(E)g(E)\, dE. \qquad (10\text{-}7)$$

Since semiconductor crystals come in various sizes, it is really more useful to calculate the *density* of electrons in the CB, given by $n = N/V$, or

$$n = \frac{1}{V} \int_{E_g}^{\infty} f_D(E)g(E)\, dE. \qquad (10\text{-}8)$$

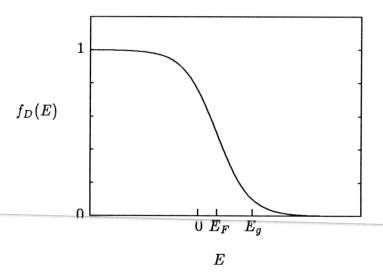

Fig. 10-3. Fermi-Dirac distribution function. The Fermi energy E_F is near the center of the gap.

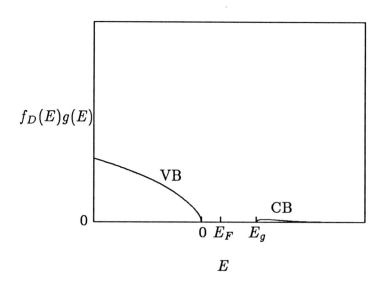

Fig. 10-4. Density of states occupied by electrons in the model semiconductor.

CHAPTER 10 SEMICONDUCTORS

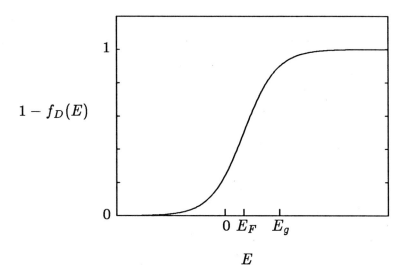

Fig. 10-5. Probability that a state is unoccupied.

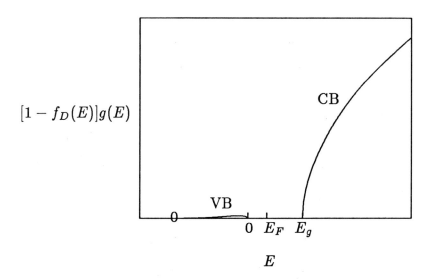

Fig. 10-6. Density of states occupied by holes in the model semiconductor.

At room temperature, we normally find that $E_g \gg k_B T$. Since E_F is approximately in the center of the gap, then $E - E_F \gg k_B T$ for all values of E over the range of the integral in Eq. (10-7). Consequently, we can use

$$f_D(E) \cong \exp\left(-\frac{E - E_F}{k_B T}\right) \tag{10-9}$$

in the integral. The integral can be now done, using $g(E)$ from Eq. (10-4), and we obtain

$$n = 2\left(\frac{m_n^* k_B T}{2\pi \hbar^2}\right)^{3/2} \exp\left(\frac{E_F - E_y}{k_B T}\right). \tag{10-10}$$

Problem 10-2. The form of $g(E)$ given in Eq. (10-4) is actually only valid near the bottom of the CB where $k \cong 0$. In Eq. (10-7), we integrate $f_D(E)g(E)$ out to $E = \infty$. How can we justify using Eq (10-4) in this integral?

Problem 10-3. Do the integral in Eq. (10-8) and obtain Eq. (10-10). You will need

$$\int_0^\infty \sqrt{x} e^{-x}\, dx = \tfrac{1}{2}\sqrt{\pi},$$

which you could have obtained from an integral table.

The density p of holes in the VB can be calculated in a similar manner:

$$p = \frac{1}{V}\int_{-\infty}^0 [1 - f_D(E)]g(E)\, dE. \tag{10-11}$$

In this case, we can use the approximation,

$$1 - f_D(E) \cong \exp\left(-\frac{E_F - E}{k_B T}\right). \tag{10-12}$$

Problem 10-4. If $E_g \gg k_B T$ and E_F is near the center of the energy gap, show that the approximation in Eq. (10-12) is valid for all values of E over the range of the integral in Eq. (10-11).

When we do the integral in Eq. (10-11), we obtain

$$p = 2\left(\frac{m_p^* k_B T}{2\pi \hbar^2}\right)^{3/2} \exp\left(-\frac{E_F}{k_B T}\right). \qquad (10\text{-}13)$$

Since the number of holes in the VB is equal to the number of electrons in the CB, we must have $n = p$. Setting Eqs. (10-10) and (10-13) equal to each other, we can solve for E_F and obtain

$$E_F = \tfrac{1}{2}E_g + \tfrac{3}{4}k_B T \ln(m_p^*/m_n^*). \qquad (10\text{-}14)$$

Problem 10-5. Set Eqs. (10-10) and (10-13) equal to each other and obtain the expression for E_F in Eq. (10-14).

Note that if $m_p^* = m_n^*$, then, from Eq. (10-14), $E_F = \tfrac{1}{2}E_g$. Even for $m_p^* \neq m_n^*$, we still have $E_F \cong \tfrac{1}{2}E_g$ since $k_B T \ll E_g$. The Fermi energy is thus near the center of the gap. Substituting Eq. (10-14) into either Eq. (10-10) or Eq. (10-13), we can find an expression for n and p:

$$n = p = 2\left(\frac{k_B T}{2\pi \hbar^2}\right)^{3/2} (m_n^* m_p^*)^{3/4} \exp\left(-\frac{E_g}{2k_B T}\right). \qquad (10\text{-}15)$$

Problem 10-6. Find the density of electrons in the CB in a crystal of pure silicon (Si) at 300 K. Repeat for 373 K. Assume that $m_n^* = m_p^* = m$, the mass of a free electron. Answer: 9.8×10^{15} m^{-3}, 9.4×10^{17} m^{-3}.

In the problem above, we see that if we raise the temperature from room temperature to that of the boiling point of water, the density n of electrons in the CB of pure silicon increases by a factor of about 100. The number of electrons that can be thermally excited to the CB is a very strong function of temperature.

10-4 Physical Picture

At this point, let us describe a physical picture of a semiconductor. Most semiconductors have the diamond or zincblende structure. In these structures, each atom has four nearest neighbors and forms covalent bonds with these neighbors. For example, Si and Ge both have the diamond structure. Compounds like GaAs, GaP, InAs, and InP are called III-V semiconductors and have the zincblende structure. Ga and In have three valence electrons, and As and P have five valence electrons (hence "III-V"). There are, on the average, four valence electrons per atom which are necessary to form four covalent bonds to each atom. Similarly, there are also II-VI semiconductors like CdS and ZnS.

Although electrons are tied up in covalent bonds between neighboring atoms, some get enough thermal energy to break away from the bond. An amount of energy at least as large as E_g is required for this to happen. These freed electrons wander throughout the crystal. The bond left behind is now deficient. It is missing an electron. The deficient bond is called a hole. An electron from a nearby bond may jump over to this deficient bond, making the bond whole but now leaving behind another deficient bond. As this process repeats itself, the deficient bond wanders throughout the crystal. Thus we get movement of holes. Occasionally a wandering electron comes near a deficient bond and gets "captured" by it. The electron and hole *recombine* and destroy each other. At thermal equilibrium, electron-hole pairs are being thermally created at the same rate that electron-hole pairs are recombining.

10-5 Doped Semiconductors

Let us now see what happens if we mix a small number of arsenic (As) atoms into a silicon (Si) crystal. These As atoms are "impurities" in the crystal. A semiconductor crystal containing such impurities is said to be **doped**. The As atoms occupy sites in the Si crystal that were formerly occupied by Si atoms when the crystal was pure. Each As atom has four Si neighbors. It forms a covalent bond to each neighbor, contributing an electron to each bond. But As has *five* valence electrons and thus has one left over, and, since the extra electron is not part of a covalent bond, it is easily removed from the As atom and can wander through the crystal. We call the state of this electron bound to the As atom an **impurity state**. Since it requires very little energy to remove this electron from this state and put it into the CB, we see that the energy level of this impurity state must be just below the bottom of the CB. The As atoms are called **donors** (they donate an electron to the CB), and the energy level of the impurity state is called the **donor level**. The donor level is at an energy E_d below the bottom of the CB (see Fig. 10-7).

Next, instead of As, let us put gallium (Ga) atoms into the Si crystal. Ga has three valence electrons. Thus, when it tries to form four covalent bonds with the neighboring Si atoms, it does not have enough electrons, and one of the bonds will be deficient. It takes very little energy for an electron from a nearby Si-Si bond to jump into this deficient bond, making the bond whole and leaving a hole behind. Since this electron comes from the VB, we see that the level of this impurity state must be just above the top of the VB. The Ga atoms are called **acceptors** (they accept electrons), and the energy level of the impurity state is called the **acceptor level**. The acceptor level is at an energy E_a above the top of the VB (see Fig. 10-8). The values of E_d and E_a for a number of different impurities in Si and Ge are given in Appendix 6.

What happens to n and p if we dope a semiconductor with

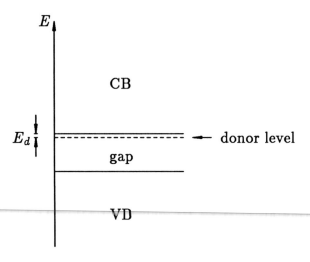

Fig. 10-7. The donor level in a semiconductor.

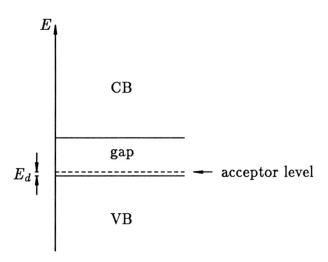

Fig. 10-8. The acceptor level in a semiconductor.

donors or acceptors? Let n_i and p_i be the densities of electrons and holes in a pure semiconductor. If we dope the crystal with donors (density N_d), almost all of these will donate electrons to the CB. (We will show this later.) If $N_d \gg n_i$, then we now have $n \cong N_d$ in the CB. With more electrons in the CB, they are more likely to meet a hole and recombine. This proceeds until most of the holes are gone. Since $p_i = n_i$, which is much less than the density of electrons $n = N_d$, these recombinations hardly affect the number of electrons at all. So we finally have $n \cong N_d$ and $p \cong 0$. This is called an **n-type semiconductor**. Similarly, if we dope the crystal with acceptors (density N_a), we get $p \cong N_a$ in the VB. These recombine with electrons in the CB until the electrons are nearly gone. This results in $p \cong N_a$ and $n \cong 0$. This is called a **p-type semiconductor**.

We can explicitly calculate n and p for a doped semiconductor. First, note that the expressions for n and p in Eqs. (10-10) and (10-13) are still valid for a doped semiconductor. The Fermi energy E_F has only shifted a bit to give either more electrons in the CB or more holes in the VB. If we take the product np, we get from Eqs. (10-10) and (10-13),

$$np = 4 \left(\frac{k_B T}{2\pi \hbar^2} \right)^3 (m_n^* m_p^*)^{3/2} \exp\left(-\frac{E_g}{k_B T} \right). \qquad (10\text{-}16)$$

Note that np is *independent* of E_F. Thus, np is a constant, independent of the doping.

In a pure semiconductor,

$$np = n_i p_i = n_i^2, \qquad (10\text{-}17)$$

since $n_i = p_i$. Consider an n-type semiconductor. We assume that all of the donors give their electrons to the CB. If $N_d \gg n_i$, then $n \cong N_d$, and from Eq. (10-17) we see that holes must recombine with electrons until $p \cong n_i^2/n \cong n_i^2/N_d$. We thus have

$$\begin{aligned} n &\cong N_d \\ p &\cong n_i^2/N_d. \end{aligned} \qquad (10\text{-}18)$$

Problem 10-7. Consider a sample of n-type silicon (Si) with $N_d = 1.00 \times 10^{21}$ m^{-3}. Find n and p at 300 K. You may use the results of Problem 10-6. Answer: 1.00×10^{21} m^{-3}, 9.6×10^{10} m^{-3}.

Problem 10-8. What fraction of total atoms in the Si crystal of Problem 10-7 are donors? Answer: 1.99×10^{-8}.

Problem 10-9. Using Eq. (10-10), calculate the Fermi energy for the doped Si crystal in Problem 10-7. Assume that $m_n^* = m$, the mass of a free electron. Answer: 0.80 eV.

Problem 10-10. Given the Fermi level in Problem 10-9, find the probability of occupation of an impurity state in that crystal if the donors are arsenic (As) atoms. (See Appendix 6.) What is the density of *occupied* impurity states? (Due to the two possible spin states of the electron, there are actually *two* impurity states associated with each impurity atom. Thus the density of impurity states is $2N_d$. This needs to be taken into account in doing this problem.) Answer: 3.0×10^{-4}, 6.0×10^{17} m^{-3}.

Problem 10-11. According to Problem 10-10, the density of occupied impurity states is 6.0×10^{17} m^{-3} while the density of occupied states in the CB is 1.00×10^{21} m^{-3}. Why are more of the states in the CB occupied than impurity states, considering that the energy of the CB is *higher* than that of the impurity states?

From Problem 10-7, we see that doping the Si crystal causes most of the holes to disappear. In a pure Si crystal, $p_i = 1.47 \times 10^{16}$ m^{-3}, and in the doped crystal, $p = 2.16 \times 10^{11}$ m^{-3}. Only a very small fraction of the electrons from the CB are needed to accomplish this. From Problem 10-8, we see that the doping level for that Si crystal is really quite small. The crystal contains only one donor for every 5×10^7 Si atoms! This small doping level causes the density of electrons to increase from $n_i = 1.00 \times 10^{16}$ m^{-3} to $n = 1.00 \times 10^{21}$ m^{-3}, a factor of 10^5!

From Problem 10-9, we see that the Fermi level in the doped Si crystal is $E_F = 0.86$ eV $= 0.77 E_g$. This is higher than the Fermi level in a pure crystal ($E_F = \frac{1}{2} E_g$) as it must be since there are now more electrons in the CB and less holes in the VB. The Fermi-Dirac distribution function shown in Fig. 10-3 has moved to the right. Using the value of E_F obtained in Problem 10-9, we can now use the Fermi-Dirac distribution function in Eq. (10-6) to calculate the probability that the impurity states are occupied. If the donors are arsenic (As) atoms, we find in Problem 10-10 that only one state in 3000 is occupied. Nearly all of the donors give their electrons to the CB.

For a p-type semiconductor, we obtain very similar results. If $N_a \gg p_i$, then

$$p \cong N_a$$
$$n \cong n_i^2/N_a. \qquad (10\text{-}19)$$

Also, we find that the Fermi energy E_F is *lowered* because now there are more holes in the VB and less electrons in the CB.

What happens if there are both donors and acceptors in the same crystal? The donors give electrons to the CB. The acceptors remove electrons from the VB, leaving holes there. There are now more electrons *and* more holes than in the pure crystal. These recombine until Eq. (10-17) is satisfied.

Problem 10-12. Consider a Si crystal doped with both donors ($N_d = 1.00 \times 10^{21}$ m^{-3}) and acceptors ($N_a = 3.00 \times 10^{21}$ m^{-3}). Find the final values of n and p at 300 K. Is the crystal n-type or p-type? You may use the results of Problem 10-6. Answer: 4.8×10^{10} m^{-3}, 2.00×10^{21} m^{-3}.

As you found in the above problem, the crystal will be n-type if $N_d > N_a$ and p-type if $N_a > N_d$. This means that we can change an n-type semiconductor to p-type by adding acceptor atoms, and we can change a p-type semiconductor to n-type by adding donor atoms.

10-6 Temperature Dependence of n

Consider the effect of temperature on n and p for an n-type semiconductor. If $N_d \gg n_i$ at room temperature, then raising the temperature a bit does not affect n very much. A few more electrons are excited from the VB into the CB, but with so many electrons already in the CB, the extra ones are not even noticed. Thus, n is temperature-independent and is entirely determined by the doping level. The semiconductor is said to be **extrinsic** here.

If we raise the temperature far enough, the thermally excited electrons will eventually become greater in number than the donated electrons. At temperatures where $n_i \gg N_d$, we find that $n \cong n_i$. The semiconductor behaves like a pure crystal at that temperature, and n increases dramatically with T. Here, the semiconductor is said to be **intrinsic** (hence the subscript i on n_i and p_i). A pure semiconductor is intrinsic at room temperature.

In Fig. 10-9 is shown the temperature dependence of n calculated for an n-type Si crystal doped with As at a density $N_d = 1.00 \times 10^{21}$ m^{-3}. The temperature dependence of n_i for a pure crystal is shown with a dashed line. We see the extrinsic region where $n \cong N_d$ and the intrinsic region where $n \cong n_i$. At very low temperatures the electrons cannot even be excited from the donor level into the CB, and n decreases to zero. This is called the **freeze-out** region.

10-7 Electrical Conductivity

Let us consider next the electrical conductivity of semiconductors. Electrons in the CB are free to wander throughout the crystal and will conduct an electric current if we apply an electric field. The electrons occupy states very near the bottom of the CB. The Fermi-Dirac distribution function is very small there so that the vast majority of states are *unoccupied*. The Pauli exclusion principle has no effect here since the electrons

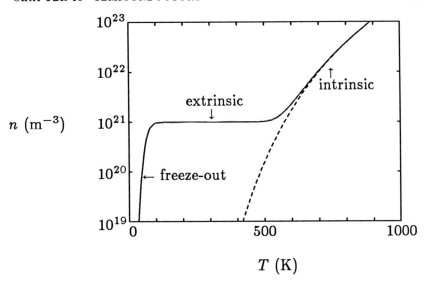

Fig. 10-9. Calculated density n of electrons in the CB of an n-type Si crystal doped with As at a density $N_d = 1.00 \times 10^{21}$ m^{-3}. The dashed line shows the density n_i of electrons in a pure Si crystal.

are not "crowded." Each electron has a wide choice of states for occupation and does not have to worry about whether or not the state it wants to go into is already occupied by another electron. The vast majority of states it has to choose from are unoccupied. These electrons behave not only like "free particles" but like "classical particles" as well.

Consequently, the behavior of these electrons is very well described by the classical model of metals in Chapter 4. The electrical conductivity of the semiconductor due to these electrons is given by Eq. (4-10),

$$\sigma_n = ne^2\tau_n/m_n^*. \qquad (10\text{-}20)$$

Since, by definition,

$$\mathbf{J} = \sigma_n \mathbf{\mathcal{E}}, \qquad (10\text{-}21)$$

and, also, from Eq. (4-7),

$$\mathbf{J} = -ne\mathbf{v}_d,$$

we have
$$\mathbf{v}_d = -\frac{\sigma_n}{ne}\boldsymbol{\mathcal{E}} = -\frac{e\tau_n}{m_n^*}\boldsymbol{\mathcal{E}}. \qquad (10\text{-}23)$$

In other words, the drift velocity $\mathbf{v}_d$ of the electrons is proportional to the electric field. We define the proportionality constant to be the **electron mobility** μ_n:

$$\mu_n = e\tau_n/m_n^*, \qquad (10\text{-}24)$$

so that we have
$$\mathbf{v}_d = -\mu_n \boldsymbol{\mathcal{E}}. \qquad (10\text{-}25)$$

Mobility is a more useful property for characterizing a semiconductor than conductivity. σ depends on n, which depends on how much we dope the material. But μ does not depend on the doping level n (as long as n is not too large). The mobility is a property of the semiconductor itself. From Eqs. (10-20) and (10-24), we find

$$\sigma_n = ne\mu_n. \qquad (10\text{-}26)$$

Holes in the VB can also conduct an electric current. Similar to electrons in the CB, these holes behave like "classical particles" with *positive* charge. The conductivity of the semiconductor due to holes is given by

$$\sigma_p = pe^2\tau_p/m_p^*. \qquad (10\text{-}27)$$

Similar to Eq. (10-24), we define a **hole mobility**,

$$\mu_p = e\tau_p/m_p^*, \qquad (10\text{-}28)$$

so that the drift velocity of holes is given by

$$\mathbf{v}_d = \mu_p \boldsymbol{\mathcal{E}}. \qquad (10\text{-}29)$$

The hole conductivity is given in terms of the mobility by

$$\sigma_p = pe\mu_p. \qquad (10\text{-}30)$$

CHAPTER 10 SEMICONDUCTORS

The electron and hole mobility for a number of semiconductors is given in Appendix 5. The total conductivity is the sum of that due to electrons and that due to holes:

$$\sigma = \sigma_n + \sigma_p. \qquad (10\text{-}31)$$

Problem 10-13. Calculate the conductivity of pure Si at 300 K due to (a) the electrons, (b) the holes, and (c) the electrons and holes together. Answer: $2.1 \times 10^{-4}/\Omega\cdot\text{m}$, $7.5 \times 10^{-5}/\Omega\cdot\text{m}$, $2.8 \times 10^{-4}/\Omega\cdot\text{m}$.

Problem 10-14. Calculate the conductivity of the n-type Si crystal in Problem 10-7 at 300 K due to (a) the electrons, (b) the holes, and (c) the electrons and holes together. Answer: $21.6/\Omega\cdot\text{m}$, $7.3 \times 10^{-10}/\Omega\cdot\text{m}$, $21.6/\Omega\cdot\text{m}$.

Problem 10-15. Calculate the mobility of conduction electrons in sodium metal (Na). Answer: 5.90×10^{-3} m^2/V·s.

Problem 10-16. A 5-mm cube of n-type germanium (Ge) at 300 K passes a current of 5.0 mA when 10.0 mV is applied between two of its parallel faces. Find the density of electrons in the conduction band. Answer: $1.6 \times 10^{21}/\text{m}^3$.

We can see in the above problems that doping a semiconductor has a large effect on its conductivity. However, compared to the conductivity of a metal ($\sigma \sim 10^7/\Omega\cdot\text{m}$), the conductivity of a semiconductor is still quite small, even when heavily doped.

The mobility of an electron or hole generally *decreases* with increasing T because of increased thermal vibrations of the lattice which scatter the moving electrons and holes. In the extrinsic region, where n or p (whichever one dominates) is temperature independent, we see from Eqs. (10-26) and (10-30) that the conductivity ought to have the same temperature dependence as μ_n or μ_p. The conductivity σ *decreases* with increasing T. In the intrinsic region at higher temperatures, just the opposite happens. Both n and p increase rapidly with T, completely overwhelming any decrease in μ_n or μ_p, and the conductivity σ increases rapidly with temperature.

10-8 Hall Effect

The Hall effect is easily observed in semiconductors. For an n-type semiconductor where the current consists of electrons, we have from the classical model, Eq. (4-17),

$$\mathcal{E}_H = -R_H \mathbf{J} \times \mathbf{B}, \qquad (10\text{-}32)$$

where the Hall coefficient R_H is given by

$$R_H = -1/ne. \qquad (10\text{-}33)$$

For a p-type semiconductor, the current consists of holes with positive charge. In Problem 9-11, we showed that the Hall coefficient is given by

$$R_H = +1/pe, \qquad (10\text{-}34)$$

which has the opposite sign as that for an n-type semiconductor. Thus, the Hall field $\mathcal{E}_H$ is in the opposite direction. The Hall effect can be used to determine if a semiconductor is n-type or p-type.

Problem 10-17. Calculate the Hall coefficient R_H for the n-type Si crystal of Problem 10-7. Answer: -6.24×10^{-3} m^3/C.

As we can see in the problem above, the Hall coefficient R_H in semiconductors is much larger than in metals ($R_H \sim 10^{-10}$ m^3/C). This is due to the much smaller density n of electrons.

For an n-type semiconductor, we find from Eqs. (10-26) and (10-33),

$$\mu_n = -\sigma R_H. \qquad (10\text{-}35)$$

CHAPTER 10 SEMICONDUCTORS

Thus we can obtain the electron mobility from experimental values of σ and R_H for an n-type semiconductor. Similarly, for a p-type semiconductor,

$$\mu_p = \sigma R_H. \qquad (10\text{-}36)$$

Problem 10-18. Consider a piece of some n-type semiconductor 5.0 cm long, 0.50 cm wide, and 1.00 mm thick. If we apply 0.25 V across the length, we find that it conducts a current of 10 mA. If we then put it in a magnetic field of 0.60 Tesla, we find that the Hall voltage across its width is 8.0 mV. Calculate the mobility of the carriers and the electron density in the CB. Answer: 0.533 m^2/V·s, 4.68×10^{21} m^{-3}.

10-9 Band Structure of Real Semiconductors

Let us next look at the band structures of some real semiconductors. First, consider the band structure of gallium arsenide (GaAs) shown in Fig. 10-10. As can be seen, the maximum energy of the VB and the minimum energy of the CB *both*

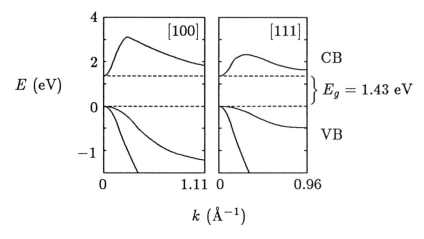

Fig. 10-10. Band structure of gallium arsenide (GaAs). See, for example, F. H. Pollak and M. Cardona, *J. Phys. Chem. Solids* **27**, 423 (1966).

occur at $k = 0$, the center of the first Brillouin zone. There is an energy gap, $E_g = 1.43$ eV, between the VB and CB. These two bands in GaAs actually each consist of several overlapping bands. We show in Fig. 10-10 only those bands which are at the bottom of the CB and at the top of the VB. These are the only bands occupied by electrons or holes. In the VB, we see *two* bands which have the same maximum energy at $k = 0$. Both of these bands are occupied by holes. Since the curvatures of these two curves near $k = 0$ are different, the effective masses of the holes in these two bands are different. We have "heavy holes" and "light holes," depending on which band the holes are in.

Problem 10-19. Which band in Fig. 10-10 contains the heavy holes and which contains the light holes?

The band structure of silicon (Si) is slightly more complicated (see Fig. 10-11). The maximum energy of the VB is at $k = 0$, as in GaAs, and there are heavy and light holes. However, the minimum energy of the CB is *not* at $k = 0$, but at $k \cong 0.99$ Å^{-1} along the [100] direction. Since there are six equivalent [100] directions in a cubic crystal, there are *six* minima of the CB in the first Brillouin zone. These are at (see Fig. 10-12)

$$\mathbf{k} = \pm(0.99 \text{ Å}^{-1})\hat{\imath}, \quad \pm(0.99 \text{ Å}^{-1})\hat{\jmath}, \quad \pm(0.99 \text{ Å}^{-1})\hat{k}.$$

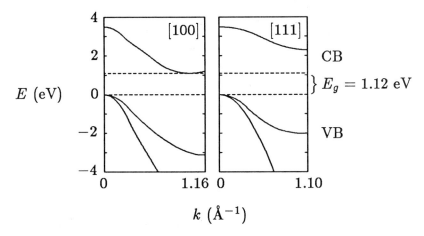

Fig. 10-11. Band structure of silicon. See, for example, M. Cardona and F. H. Pollak, *Phys. Rev.* **142**, 530 (1966).

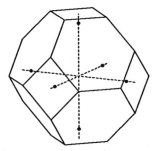

Fig. 10-12. The six minima in the CB of silicon.

CHAPTER 11

p-n JUNCTIONS IN SEMICONDUCTORS

11-1 The Junction

In this chapter, we will discuss the *p-n* junction in a semiconductor. These junctions play an important role in the operation of virtually all semiconductor devices.

Consider a semiconductor crystal which has been doped with donors in one region and with acceptors in another region. Part of the crystal is *n*-type, and part is *p*-type. The region of the crystal where the two types of doping meet each other is called a *p-n* junction. As we move across this junction, the crystal changes from *p*-type to *n*-type.

Let $x = 0$ be the position of the junction. Let the crystal be *n*-type for $x > 0$, and let the density N_d of donors be uniform throughout the region, even up to the junction. We will call this the "*n*-side" of the junction. Let the crystal be *p*-type for $x < 0$, and let the density N_a of acceptors be uniform throughout this region. We will call this the "*p*-side" of the junction. As we cross the junction at $x = 0$, we will assume that the doping changes abruptly from donors (density N_d) to acceptors (density N_a), as shown in Fig. 11-1. This kind of junction is called an **abrupt junction** (or step junction). (In practice, no junction can be made infinitely abrupt. A real crystal changes gradually from *p*-type to *n*-type across the junction.)

11-2 Diffusion of Electrons and Holes

To see what happens to the electrons and holes near the junction, let us assume that such a junction has just been put together and that the electrons and holes exist initially on each side of the junction just as in the pure *n*-type and *p*-type crystals alone (see Fig. 11-2). In the CB, there are more electrons where $x > 0$ than where $x < 0$. In the VB there are more holes where $x < 0$ than where $x > 0$.

The situation is very similar to a container of two different kinds of gases separated by a wall down the center

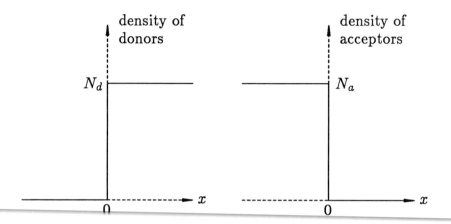

Fig. 11-1. The density of donors and acceptors at an abrupt p-n junction.

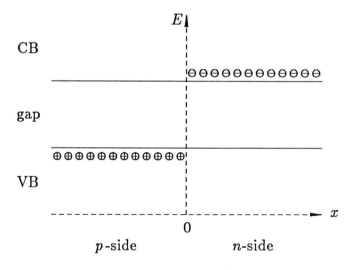

Fig. 11-2. Initial positions of electrons and holes in a p-n junction. The circles with minus signs represent electrons, and the circles with plus signs represent holes.

(see Fig. 11-3). If we suddenly remove the wall, the gases start to mix. This happens because the velocities of the gas molecules are randomly distributed. Near the wall there are some molecules which travel in a direction toward the wall. With the wall in place, they merely bounce off the wall and stay on their side of the container. With the wall removed, though, they enter the other half of the container and mix with the gas molecules of the other type. Some of these molecules eventually return to their own side of the container, but as long as there is a higher density of that type of molecule on their own side of the container than on the other, more molecules will move to the other side than will come back. This will continue until there is an equal density of each type of gas on each side of the container and the gases are thoroughly mixed. This process of mixing is called **diffusion**.

Diffusion also occurs across the p-n junction. Since there are more electrons in the CB for $x > 0$, electrons diffuse across the junction into the CB on the p-side. Similarly, holes diffuse across the junction into the VB on the n-side. Unlike the mixing of the gases, though, the diffusion of electrons and holes does not proceed until they are thoroughly mixed throughout the entire crystal. Electrons and holes carry charge. As the electrons diffuse across the junction, the region near the junction on the p-side becomes *negatively* charged and repels the rest of the electrons on the n-side, preventing further diffusion. Similarly, the diffusion of holes across the junction causes the n-side to become *positively* charged near the junction. This

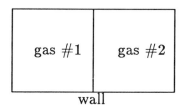

Fig. 11-3. Two kinds of gases separated by a wall.

charge repels the rest of the holes on the p-side and prevents further diffusion.

11-3 Electric Field and Contact Potential

The excess negative charge on the p-side and the excess positive charge on the n-side give rise to an electric field $\mathscr{E}$ at the junction, pointing from the n-side to the p-side of the junction (see Fig. 11-4). This electric field produces a force on the electrons in the $+x$ direction, thus preventing any more of them from diffusing from the n-side to the p-side of the junction. Also, this electric field produces a force on the holes in the $-x$ direction, preventing any more of them from diffusing from the p-side to the n-side of the junction. (The force on the electrons and holes are in opposite directions because of their opposite charge.)

There is thus a potential difference between the p-side and n-side of the junction. The potential energy of the electrons is higher for $x < 0$ than for $x > 0$. The electrons prefer to stay on the n-side, where their potential energy is lowest. Similarly, the potential energy of the holes is higher for $x > 0$ than for $x < 0$, and they prefer to stay on the p-side, where their potential energy is lowest. We draw a diagram of the energy across the junction in Fig. 11-5. Note that this diagram actually shows the energy of the *electrons* which occupy the states. The energy of the holes which occupy the states is opposite in sign because of their opposite charge. Electrons seek states lowest in energy. These are *downward* in the diagram. Holes also seek states lowest in energy, but for them, these are *upward* in the diagram. Thus, electrons cannot enter the p-side because it is "uphill." Holes cannot enter the n-side because it is "downhill." Holes prefer to go "uphill." We define the potential energy difference across the junction to be $e\phi$, where ϕ is called the **contact potential** (or built-in voltage).

11-4 Depletion Layer

Something else also happens near the junction. The excess electrons in the CB on the p-side cannot really exist there in

CHAPTER 11 p-n JUNCTIONS IN SEMICONDUCTORS   219

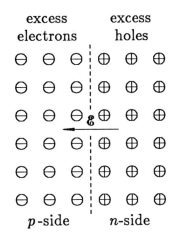

Fig. 11-4. Electric field created by excess charge at the junction.

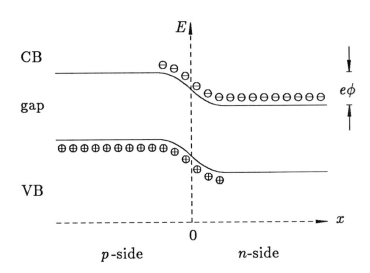

Fig. 11-5. Diffusion of electrons and holes across the junction causes a potential energy difference.

the presence of all those holes. The electrons and holes recombine. The same thing happens on the n-side near the junction. This recombination of electrons and holes greatly reduces the number of free carriers (electrons and holes) near the junction. The region where this happens is called the **depletion layer**.

Note that even though there may be very few free charges in the depletion layer, the p-side of the junction still has excess negative charge, and the n-side still has excess positive charge. (For this reason, the depletion layer is also called the space-charge region.) This must be so since the recombination of an electron and a hole does not change the total charge. In the depletion layer, the excess negative charge on the p-side comes from the acceptor atoms which each have an extra electron, and the excess positive charge on the n-side comes from the donor atoms which each have given up an electron. The electric field in the depletion layer prevents electrons and holes from entering (see Fig. 11-6).

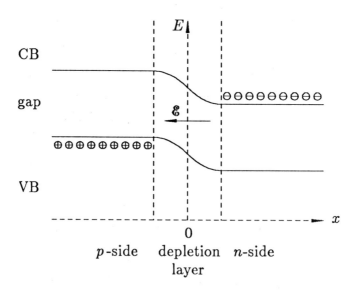

Fig. 11-6. Depletion layer at a p-n junction.

11-5 Fermi Level: Calculation of Contact Potential

This situation can be described in terms of the Fermi level. When the crystal is in equilibrium, the Fermi energy must have the same value everywhere in the crystal. To see why this must be so, consider the Fermi-Dirac distribution function $f_D(E)$. If two regions of the crystal had different Fermi energies E_F, then $f_D(E)$ would also be different in each region. [Remember that $f_D(E)$ depends on E_F.] Consider some states at a certain energy E in both regions. A different value of $f_D(E)$ means a different occupancy of states. The states at E are more filled with electrons in one region than in the other. Since electrons can freely move across the crystal to states of the same energy, this situation would cause electrons in the region of greater $f_D(E)$ to diffuse to the region of lesser $f_D(E)$. Since "equilibrium" means no net flow of electrons, we thus cannot have a state of equilibrium if $f_D(E)$ is not the same everywhere in the crystal. Otherwise the electrons would flow until $f_D(E)$ is the same everywhere. In equilibrium, $f_D(E)$ *will* be the same everywhere, and hence E_F will have the same value everywhere.

Let us apply this concept to the p-n junction. For $x \gg 0$, far from the junction on the n-side, we have $n \cong N_d$, and we can solve for the Fermi energy, using Eqs. (10-10) and (10-15),

$$E_F = \tfrac{1}{2}E_g + k_B T \ln\left[(m_p^*/m_n^*)^{3/4}(N_d/n_i)\right]. \qquad (11\text{-}1)$$

Problem 11-1. Using Eqs. (10-10) and (10-15), solve for the Fermi energy in an n-type semiconductor and obtain Eq. (11-1).

Since $N_d \gg n_i$, we see from Eq. (11-1) that $E_F > \tfrac{1}{2}E_g$. The Fermi level is more than half way across the gap, as we would expect. The energy in Eq. (11-1) is referenced to the top of the VB where we chose $E = 0$ in Chapter 10. We see in Fig. 11-6 that the energy at the top of the VB is *different* on the two sides of the junction and therefore cannot be zero on both sides simultaneously. We must be careful what we call $E = 0$. Let us no longer call the energy zero at the top of the VB. Instead, let E_{vn} be the energy at the top of the VB on the n-side, far from the junction, and let E_{vp} be the energy at the top of the VB on the p-side, far from the junction. Then, Eq. (11-1) should be written

$$E_F = E_{vn} + \tfrac{1}{2} E_g + k_B T \ln \left[(m_p^*/m_n^*)^{3/4} (N_d/n_i) \right]. \quad (11\text{-}2)$$

For $x \ll 0$, far from the junction on the p-side, we similarly have

$$E_F = E_{vp} + \tfrac{1}{2} E_g - k_B T \ln \left[(m_n^*/m_p^*)^{3/4} (N_a/n_i) \right]. \quad (11\text{-}3)$$

Problem 11-2. Using Eqs. (10-13) and (10-15), solve for the Fermi energy in a p-type semiconductor and obtain Eq. (11-3).

Since E_F must be the same on both sides of the junction, we obtain from Eqs. (11-2) and (11-3),

$$E_{vp} - E_{vn} = k_B T \ln(N_d N_a / n_i^2). \quad (11\text{-}4)$$

Problem 11-3. Set Eqs. (11-2) and (11-3) equal to each other and obtain Eq. (11-4).

From Eq. (11-4) we see that the energy at the top of the VB on the p-side is *greater* than the energy at the top of the VB on the n-side ($E_{vp} > E_{vn}$), in agreement with Fig. 11-5. We defined this potential difference to be $e\phi$:

$$e\phi = E_{vp} - E_{vn}. \quad (11\text{-}5)$$

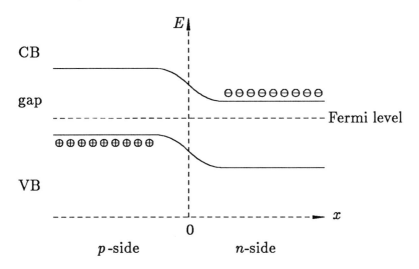

Fig. 11-7. Fermi level across a p-n junction.

Thus,
$$\phi = (k_B T/e) \ln(N_d N_a / n_i^2). \quad (11\text{-}6)$$

In Fig. 11-7 is the energy diagram of the junction with the Fermi level shown as the dashed line. Note in the figure that the Fermi level is the same everywhere. However, the energy of the *states* changes across the junction. On the n-side, the CB is closer to the Fermi level than the VB, resulting in more electrons in the CB than holes in the VB. On the p-side, the VB is closer to the Fermi level than the CB, resulting in more holes in the VB than electrons in the CB.

Problem 11-4. Consider a p-n junction in silicon (Si) with $N_d = 1.00 \times 10^{21}$ m^{-3} in the n-type Si and $N_a = 1.00 \times 10^{21}$ m^{-3} in the p-type Si. Calculate the contact potential ϕ at 300 K. You may use the result of Problem 10-6. Answer: 0.596 V.

From the above problem, we see that the energy difference across the junction is 0.596 eV, a sizeable fraction of the total gap width ($E_g = 1.12$ eV in Si).

11-6 Width of Depletion Layer

Knowing the contact potential ϕ, we can also estimate the width of the depletion layer. For simplicity, let us assume that $N_a = N_d$ and that the boundary of the depletion layer is sharp. Within the depletion layer, there are *no* free carriers, and outside the depletion layer, the concentration of free carriers is the same as that far from the junction. Then, on the n-side of the junction, $n = 0$ inside the depletion layer, and $n = N_d$ outside. Similarly, on the p-side, $p = 0$ inside the depletion layer, and $p = N_a$ outside (see Fig. 11-8). (In a real semiconductor, the transition from 0 to N_d and from 0 to N_a is actually gradual.)

Outside the depletion layer, the charge of the electrons and holes is balanced by the charge of the impurity atoms, and the net charge density is zero. Inside the depletion layer the charge of the impurity atoms is no longer canceled (there are no electrons or holes), and the charge density is $\rho = +eN_d$ on the n-side and $\rho = -eN_a$ on the p-side (see Fig. 11-9).

To conserve charge neutrality in the crystal, there must be as much negative charge as positive charge at the junction. If $N_d = N_a$, then we see that the depletion layer must be *centered* on $x = 0$. Defining x_d to be the *total* width of the depletion layer, the depletion layer extends from $x = -\frac{1}{2}x_d$ to $x = \frac{1}{2}x_d$.

We can find the electric field in the depletion layer by using Gauss's Law,

$$\oint \boldsymbol{\mathcal{E}} \cdot d\mathbf{S} = Q/\epsilon. \tag{11-7}$$

The left-hand side of the equation is the surface integral over some chosen volume. The quantity Q is the *net* charge contained within that volume. The constant ϵ is the permittivity of the semiconductor and is related to the permittivity ϵ_0 of vacuum by the equation,

$$\epsilon = \epsilon_r \epsilon_0, \tag{11-8}$$

where ϵ_r is the *relative* permittivity (or **dielectric constant**) of the semiconductor ($\epsilon_r = 12$ in Si, and $\epsilon_r = 16$ in Ge). For the p-n junction, we know by symmetry that $\boldsymbol{\mathcal{E}}$ points along

CHAPTER 11 p-n JUNCTIONS IN SEMICONDUCTORS   225

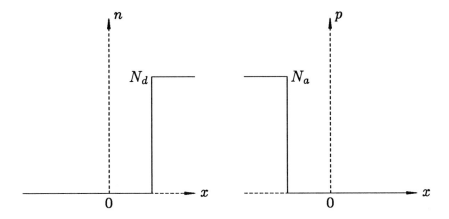

Fig. 11-8. Density of electrons and holes near the junction.

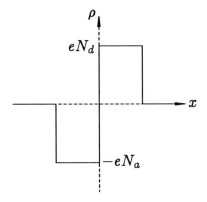

Fig. 11-9. Charge density ρ near the junction.

the positive or negative x direction and has no components in the y and z directions:

$$\mathbf{\mathcal{E}} = \mathcal{E}\,\hat{\imath}. \tag{11-9}$$

With this form of $\mathbf{\mathcal{E}}$, we can write Gauss's Law in a more convenient form:

$$\frac{d\mathcal{E}}{dx} = \frac{\rho}{\epsilon_r \epsilon_0}. \tag{11-10}$$

Problem 11-5 Show that Gauss's Law takes the form of Eq. (11-10) if the electric field only has an x-component, as in Eq. (11-9). Hint: do the surface integral over a small cube of sides dx, dy, dz.

Since the *total* charge at the junction is zero, the electric field must be zero at an infinite distance away (either in the $+x$ or $-x$ direction). Thus we can obtain $\mathcal{E}(x)$ by integrating Eq. (11-10) from $-\infty$ to x:

$$\mathcal{E} = \frac{1}{\epsilon_r \epsilon_0} \int_{-\infty}^{x} \rho\, dx. \tag{11-11}$$

Using the function shown in Fig. 11-9 for ρ, we can do the integral and obtain

$$\mathcal{E} = \begin{cases} -(eN_d/\epsilon_r\epsilon_0)(\tfrac{1}{2}x_d + x), & -\tfrac{1}{2}x_d < x < 0, \\ -(eN_d/\epsilon_r\epsilon_0)(\tfrac{1}{2}x_d - x), & 0 < x < \tfrac{1}{2}x_d, \\ 0, & \text{otherwise.} \end{cases} \tag{11-12}$$

Remember that we are treating the case where $N_a = N_d$. We plot $\mathcal{E}$ in Fig. 11-10. We see that the magnitude of the electric field has a maximum value at $x = 0$ given by

$$\mathcal{E}_{\max} = eN_d x_d / 2\epsilon_r \epsilon_0. \tag{11-13}$$

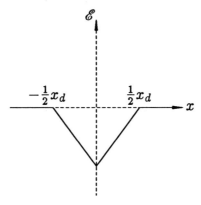

Fig. 11-10. The electric field $\mathscr{E}$ near the junction.

Problem 11-6. Integrate Eq. (11-11) and obtain the electric field given in Eq. (11-12).

The difference in potential energy across the junction is given by the integral,

$$\Delta U = -\int_{-\infty}^{\infty} F \, dx, \qquad (11\text{-}14)$$

where F is the force on an electron. Since $F = -e\mathscr{E}$ in the depletion layer,

$$\Delta U = e \int_{-x_d/2}^{x_d/2} \mathscr{E} \, dx, \qquad (11\text{-}15)$$

which is simply the area under the curve in Fig. 11-10 multiplied by e.

$$\Delta U = -e^2 N_d x_d^2 / 4\epsilon_r \epsilon_0. \qquad (11\text{-}16)$$

Problem 11-7. Find ΔU from the area under the curve in Fig. 11-10.

The potential energy of the electron decreases as we go from the p-side to the n-side of the junction, consistent with Fig. 11-7. Thus, $\Delta U = -e\phi$, and we have

$$\phi = eN_d x_d^2 / 4\epsilon_r \epsilon_0. \tag{11-17}$$

Solving for x_d, we obtain

$$x_d = \sqrt{4\epsilon_r \epsilon_0 \phi / eN_d}. \tag{11-18}$$

Substituting this into Eq. (11-13), we find that at the center of the depletion layer, the magnitude of the electric field is given by

$$\mathscr{E}_{max} = \sqrt{eN_d \phi / \epsilon_r \epsilon_0}. \tag{11-19}$$

Problem 11-8. Obtain Eq. (11-19) by substituting Eq. (11-18) into Eq. (11-13).

Problem 11-9. Find the width of the depletion layer for the p-n junction in Problem 11-4 ($\epsilon_r = 12$ in Si). Answer: 1.26 μm.

Problem 11-10. Find the magnitude of the electric field at the center of the depletion layer for the p-n junction in Problem 11-4. Answer: 9.48×10^5 V/m.

11-7 Currents across Junction in Equilibrium

Let us next examine more closely the movement of electrons and holes across the junction. First consider the electrons. Some electrons in the CB on the n-side get enough thermal energy to get over the energy barrier $e\phi$ and into the p-side. As soon as these get over there, they recombine with holes in the VB. This current of electrons across the junction is called the **recombination current** I_{nr0} (see Fig. 11-11a). Note that the current is in the opposite direction as the movement of the electrons since the electrons have negative charge. In addition to the recombination current, some electrons in the

(a) Recombination current

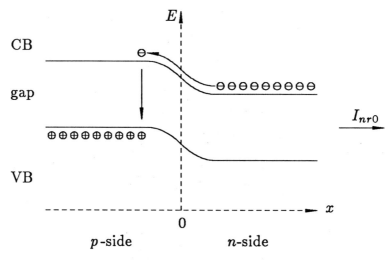

(b) Generation current

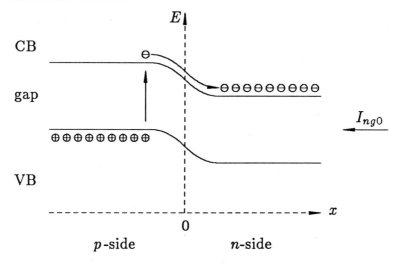

Fig. 11-11. The current of electrons across the junction.

(a) Recombination current

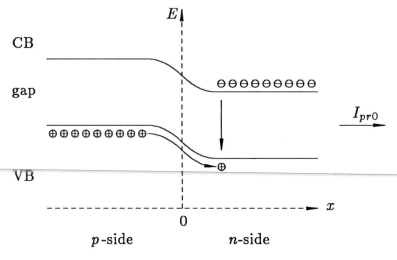

(b) Generation current

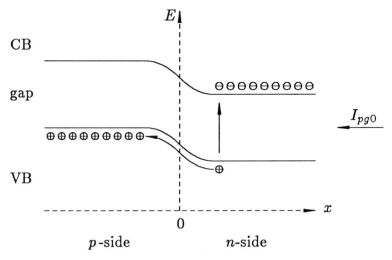

Fig. 11-12. The current of holes across the junction.

VB on the p-side are thermally excited into the CB, leaving an extra hole behind. Once in the CB, the electron easily moves across the junction to the n-side. This current is called the **generation current** I_{ng0} (see Fig. 11-11b). Since at equilibrium, the *net* current of electrons across the junction must be zero, the two currents must be equal to each other in magnitude (they are in opposite directions):

$$I_{nr0} = I_{ng0}. \tag{11-20}$$

Similarly, there is also for the holes a recombination current I_{pr0} and generation current I_{pg0} (see Fig. 11-12), and we have

$$I_{pr0} = I_{pg0}. \tag{11-21}$$

11-8 Biased Junctions

Let us next apply an electric voltage across the junction so that a net current flows. This is called a **biased junction**. Depending on the polarity of the applied voltage, this either increases or decreases the potential energy of an electron on one side of the junction with respect to the that of an electron on the other side. The energy difference across the junction is now either greater than or less than $e\phi$.

Let us apply a voltage V_a across the junction such that the p-side is positive with respect to the n-side. (We connect the positive terminal of a battery to the p-side and the negative terminal to the n-side.) This is called a **forward biased junction**. The potential energy of the electrons on the p-side is now less than before. The positive terminal of the battery "attracts" the electrons over into the p-side. This lowers the potential barrier at the junction by an amount eV (see Fig. 11-13). The energy difference across the junction is now $e(\phi - V_a)$. Since the energy barrier is smaller, more electrons should have enough thermal energy to get into the p-side, and the recombination current should increase.

We can calculate the density of electrons that have enough energy to get over the barrier. We denote this density by $n(V_a)$.

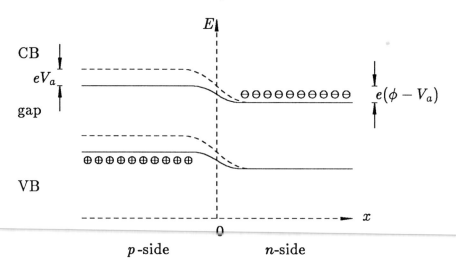

Fig. 11-13. A forward biased junction.

If we call E_{vn} the energy at the top of the VB on the n-side, then the energy E_{cp} at the bottom of the CB on the p-side is given by

$$E_{cp} = E_{vn} + E_g + e(\phi - V_a). \tag{11-22}$$

Electrons on the n-side must have an energy greater than E_{cp} to cross the junction into the p-side. These are the electrons which occupy *states* on the n-side with energy greater than E_{cp}. We can find the density $n(V_a)$ of these electrons from

$$n(V_a) = \frac{1}{V} \int_{E_{cp}}^{\infty} f_D(E) g(E)\, dE, \tag{11-23}$$

where $g(E)$ is the density of states on the n-side. This equation is very similar to Eq. (10-8) for the density n of *all* the electrons in the CB. We can integrate this equation, just as we did for Eq. (10-8), and obtain, similar to Eq. (10-10),

$$n(V_a) = 2 \left(\frac{m_n^* k_B T}{2\pi \hbar^2} \right)^{3/2} \exp\left(\frac{E_F - E_{cp}}{k_B T} \right), \tag{11-24}$$

or, using Eq. (11-22),

$$n(V_a) = 2 \left(\frac{m_n^* k_B T}{2\pi \hbar^2}\right)^{3/2} \exp\left(\frac{E_F - E_{vn} - E_g - e(\phi - V_a)}{k_B T}\right). \tag{11-25}$$

This can be expressed as

$$n(V_a) = n(0) \exp(eV_a/k_B T), \tag{11-26}$$

where $n(0)$ is the density of electrons with enough energy to cross the barrier when no voltage is applied ($V_a = 0$). We see that the number of electrons that can cross the junction into the p-side increase exponentially with the applied voltage V_a.

The electron recombination current I_{nr} is proportional to the number of electrons that can get over the barrier. The recombination current increases the same way that $n(V_a)$ does, and we have from Eq. (11-26),

$$I_{nr} = I_{nr0} \exp(eV_a/k_B T), \tag{11-27}$$

where I_{nr0} is the recombination current when the applied voltage is zero. The holes on the p-side also find it easier to get over the barrier into the n-side. Using the same line of reasoning as we did for electrons, we find that the hole recombination current I_{pr} is given by

$$I_{pr} = I_{pr0} \exp(eV_a/k_B T). \tag{11-28}$$

The electron generation current I_{ng} is not affected by the size of the energy barrier at the junction. Electrons are thermally generated on the p-side at the same rate as before, and these electrons find it just as easy to cross the junction into the n-side, no matter how large the barrier is, because, for them, it is a "downhill" ride across the junction. Thus, we have

$$I_{ng} = I_{ng0}. \tag{11-29}$$

Similarly, for holes,

$$I_{pg} = I_{pg0}. \tag{11-30}$$

The total current across the junction is given by

$$I = I_{nr} - I_{ng} + I_{pr} - I_{pg}. \qquad (11\text{-}31)$$

Combining Eqs. (11-20), (11-21), and (11-27) through (11-31), we finally obtain

$$I = I_0 \left[\exp(eV_a/k_BT) - 1\right], \qquad (11\text{-}32)$$

where

$$I_0 = I_{ng0} + I_{pg0}. \qquad (11\text{-}33)$$

As we can see, the current increases exponentially with V_a.

Problem 11-11. Use Eqs. (11-20), (11-21), and (11-27) through (11-31) to obtain Eq. (11-32).

If we reverse the polarity of V_a, the energy barrier is *increased* instead of decreased (see Fig. 11-14). In this case, we say the junction is **reverse biased**. All of the equations we developed for the forward biased junction hold for the reverse biased junction also. We just use a negative value for V_a. With a larger energy barrier, the recombination currents decrease, as can be seen in Eqs. (11-27) and (11-28) if $V_a < 0$. Fewer electrons and holes have enough thermal energy to get over the barrier. On the other hand, the generation currents are still unaffected by the barrier height, and they are given by Eqs. (11-29) and (11-30). For $V \ll -k_BT$, the recombination current becomes negligibly small, and the net current is dominated by the generation current and is equal to I_0 given by Eq. (11-33).

We plot in Fig. 11-15 the current I as a function of applied voltage. The junction can conduct a large amount of current in one direction (when $V_a > 0$) but very little in the other direction (when $V_a < 0$). Such a device is called a **rectifier** or **diode**. A typical value for I_0 is of the order of 10 μA.

CHAPTER 11 p-n JUNCTIONS IN SEMICONDUCTORS   235

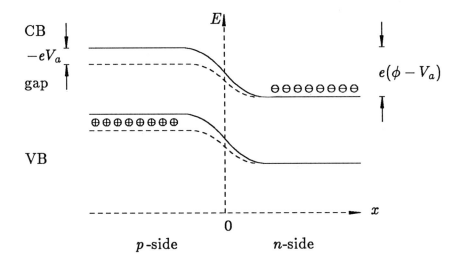

Fig. 11-14. A reverse biased junction.

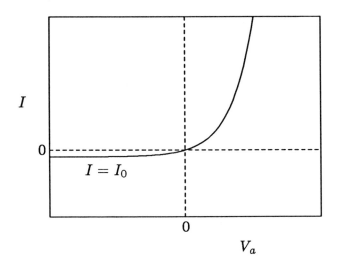

Fig. 11-15. Current I across a p-n junction as a function of applied voltage V_a.

Problem 11-12. (a) If $I_0 = 10.0$ μA for some p-n junction, find the current at 300 K if it is forward biased with 0.2 V. (b) Find the current if it is reverse biased with 0.2 V. Answer: 23.0 mA, -10.0 μA.

What happens to the Fermi level when the p-n junction is conducting a current? The electrons and holes are *not* in a state of equilibrium (they are flowing). Thus, the Fermi level is not the same on both sides of the junction. When the junction is forward biased, for example, the Fermi energy on the p-side is lower than that on the n-side, and current flows, trying to make the Fermi energy equal on each side (see Fig. 11-16a). When the junction is reverse biased, the opposite happens (see Fig. 11-16b).

Since the barrier height of a junction changes when it is biased, the width x_d of the depletion layer must also change. We can obtain the new value of x_d by simply substituting the new barrier height $\phi - V$ for ϕ in Eq. (11-18):

$$x_d = \sqrt{4\epsilon_r\epsilon_0(\phi - V_a)/eN_d}. \tag{11-34}$$

Substituting this into Eq. (11-19) gives us the magnitude of the electric field at the center of the depletion layer:

$$\mathscr{E}_{max} = \sqrt{eN_d(\phi - V_a)/\epsilon_r\epsilon_0}. \tag{11-35}$$

We see that in a forward biased junction ($V_a > 0$), the magnitude of the electric field is decreased. This happens because the battery connected across the junction produces an electric field in the $+x$ direction, opposite to the direction of the electric field already there. The width x_d of the depletion layer also decreases because less charge is needed there to produce a smaller electric field.

In a reverse biased junction ($V_a < 0$), the battery produces a field in the $-x$ direction, which is in the *same* direction as the

(a) Forward biased

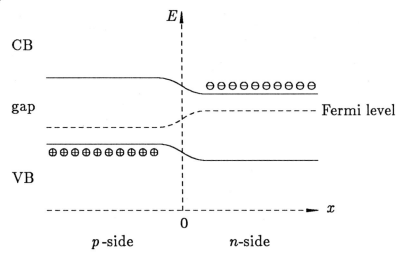

(b) Reverse biased

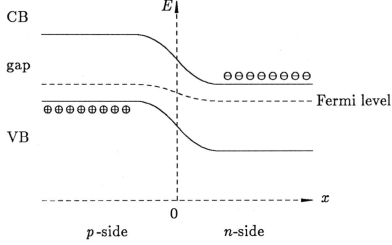

Fig. 11-16. The Fermi level when current is flowing across a junction.

electric field already there. Thus, the magnitude of the field *increases*. The width x_d of the depletion layer also increases because more charge is needed there to produce a larger field. We can see this change in width illustrated in Fig. 11-16.

Problem 11-13. Consider the junction in Problem 11-4. If we reverse-bias this junction with 10.0 V, find the width of the depletion layer ($\epsilon_r=12$ in Si). Answer: 5.30 μm.

11-9 Capacitance of Junction

A *p-n* junction has capacitance. As we can see in Fig. 11-9, there is a positive charge on one side of the junction and a negative charge on the other side, very similar to the charges on a parallel plate capacitor. Also, since the width of the depletion layer changes with the bias voltage V_a, so does the charge Q.

Consider a junction of area A. If the depletion layer extends a distance $x_d/2$ into the *n*-side of the junction, then the volume of the depletion layer there is equal to $x_d A/2$. The charge density in the depletion layer is equal to eN_d. Therefore, the charge Q on the *n*-side is given by

$$Q = (x_d A/2)(eN_d). \qquad (11\text{-}36)$$

The charge on the *p*-side, of course, has the same magnitude but opposite sign. Using Eq. (11-34), we find

$$Q = A\sqrt{eN_d \epsilon_r \epsilon_0 (\phi - V_a)}. \qquad (11\text{-}37)$$

The capacitance C is defined to be

$$C = dQ/dV. \qquad (11\text{-}38)$$

Using Eq. (11-37), we finally obtain

$$C = A\sqrt{eN_d \epsilon_r \epsilon_0 / 4(\phi - V_a)}. \qquad (11\text{-}39)$$

Problem 11-14. Consider the junction in Problem 11-4. Let the area of the junction be 1.00 mm². If we reverse-bias this junction with 10.0 V, find its capacitance. ($\epsilon_r = 12$ in Si.) Answer: 20 pF.

You may notice that the capacitance depends on the voltage (unlike parallel plate capacitors). The capacitance decreases as the reverse bias is increased. A veractor diode is a device which uses a reverse biased p-n junction as a voltage-controlled capacitor.

CHAPTER 12

SEMICONDUCTOR DEVICES

12-1 Diode

In this chapter, we will describe the operation of a number of semiconductor devices. First, let us consider the **diode** or **rectifier**. A diode is simply a single p-n junction. A metal wire is connected to each side of the junction, as shown schematically in Fig. 12-1.

If we apply a positive voltage to the p-side with respect to the n-side, the junction is forward biased, and a large amount of current will flow. If we apply a positive voltage to the n-side with respect to the p-side, the junction is reverse biased, and very little current will flow. A diode only allows current to flow in one direction, from the p-side to the n-side. The electronic symbol for the diode is also shown in Fig. 12-1.

Let us trace the current through the diode when it is forward biased. The current is dominated by the electron and hole recombination currents, I_{nr} and I_{pr} (see Fig. 12-2). Electrons flow from the n-side to the p-side in the CB, and holes flow from the p-side to the n-side in the VB. The electrons which have crossed the junction recombine with holes, and the holes which have crossed the junction recombine with electrons.

Thus, electrons in the CB of the n-side are being depleted in two ways. (1) They are recombining with holes that have

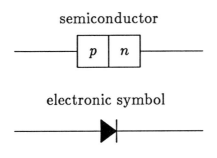

Fig. 12-1. Diode.

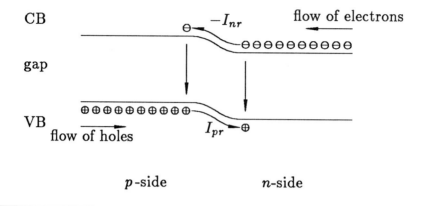

Fig. 12-2. Electron and hole recombination currents in a forward biased p-n junction.

crossed the junction in the VB, and (2) they are leaving the n-side and going into the CB on the p-side. To replenish the electrons, there must be a net flow of electrons in the n-side toward the junction. These electrons come from the wire which connects the n-side to the negative terminal of the battery.

Similarly, holes in the VB on the p-side are being depleted in two ways. (1) They are recombining with electrons that have crossed the junction in the CB, and (2) they are leaving the p-side and going into the VB on the n-side. To replenish these holes, there must be a net flow of holes in the p-side toward the junction. These holes are produced in the region where the wire is connected to the p-side. The wire removes electrons from the VB of the p-side, leaving holes, and carries the electrons to the positive terminal of the battery.

Thus, in a forward biased diode, electrons are flowing toward the junction on the n-side, and holes are flowing toward the junction on the p-side. At the junction they recombine and destroy each other. (In practice, diodes are quite small so that most of the holes and electrons do not recombine by the time the reach the metal leads.)

In a reverse biased diode, very little current flows. However, if the voltage across a reverse biased junction becomes too great, the current will suddenly increase to a very large value. This is called **reverse breakdown**. There are two mechanisms which can cause this: avalanche breakdown and Zener breakdown. In **avalanche breakdown**, the electric field in the depletion layer becomes so great that any electrons there are greatly accelerated and acquire large kinetic energy. These electrons collide with atoms, knocking electrons out of the bonds (from the VB to the CB). These electrons in turn acquire enough energy to knock out other electrons. The number of free electrons quickly multiplies, much like an avalanche. The large number of electrons in the CB and the corresponding large number of holes in the VB cause the current to increase dramatically.

Zener breakdown occurs for a very different reason. When V is large and negative, it is possible for the CB on the n-side to have *lower* energy than the VB on the p-side (see Fig. 12-3a). Let us redraw this diagram, showing in the VB the occupation of states by *electrons*, rather than holes (see Fig. 12-3b). On the n-side, the states in the CB are mostly *unoccupied*. On the p-side, the states in the VB are mostly *occupied* (by electrons). The electrons on the p-side would rapidly flow into the unoccupied states on the n-side since they are at the same energy, except that there is a region at the junction where *no* states exist at those energies. Ordinarily, to get to these unoccupied states, an electron must first get into the CB and *then* cross the junction. This is the electron generation current we discussed in Chapter 11 and is very small. If the forbidden region is very narrow, however, the electrons may **tunnel** across the junction (see Fig. 12-3b), giving rise to a current. This is what happens in Zener breakdown. For Zener breakdown to occur, we need a very narrow depletion layer. This is usually achieved by doping with a high density of impurities (greater than 10^{24} m^{-3}).

Problem 12-1. Consider a p-n junction in a silicon crystal (Si) with $N_d = 1.00 \times 10^{24}$ m^{-3} in the n-type Si and $N_a = 1.00 \times 10^{24}$ m^{-3} in the p-type Si. If the junction is

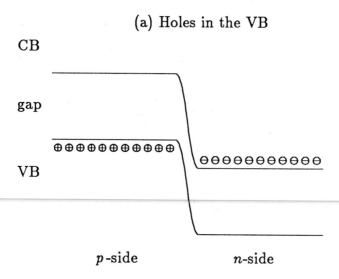

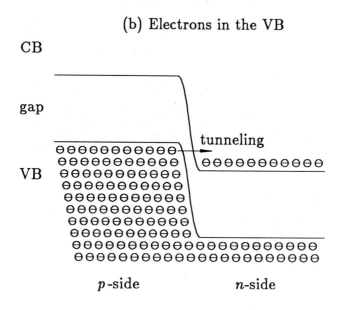

Fig. 12-3. Zener breakdown.

CHAPTER 12 SEMICONDUCTOR DEVICES   245

unbiased at 300 K, what is the width of the depletion layer? See Problem 11-9. Answer: 500 Å.

Reverse breakdown is used in a device called a **Zener diode**. The voltage at which reverse breakdown occurs can be controlled by the design of the diode. By limiting the current, Zener diodes can be operated in the reverse breakdown condition and are very useful for voltage regulation in a circuit. In Zener diodes, reverse breakdown may occur by either avalanche breakdown or Zener breakdown or both.

12-2 Bipolar Junction Transistor

One of the most important inventions of our modern world is the **transistor**. This semiconductor device was invented by J. Bardeen and W. H. Brattain in 1949 at Bell Telephone Laboratories. Their first working model was essentially a bipolar junction transistor. This kind of transistor is still used today.

A bipolar junction transistor consists of three layers of doped semiconductor. There are two ways to construct it: a layer of p-type semiconductor sandwiched between two layers of n-type semiconductor (called an n-p-n transistor) or a layer of n-type semiconductor sandwiched between two layers of p-type semiconductor (called a p-n-p transistor), as shown in Fig. 12-4. The middle layer is called the **base** and the two outer layers are called the **emitter** and the **collector**. A wire is connected to each layer, as shown in the figure. You should note that the entire transistor is a *single* crystal which has simply been doped differently in the three layers.

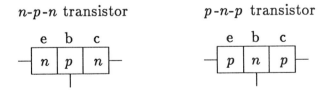

Fig. 12-4. Two kinds of bipolar junction transistors showing the emitter (e), base (b), and collector (c).

We will consider here the *n*-*p*-*n* transistor. With no applied voltage, the energy diagram looks like that in Fig. 12-5. The base is made of *p*-type semiconductor, and the emitter and collector are each made of *n*-type semiconductor. As can be seen in Fig. 12-5, the free charge carriers consist of mainly holes in the base, and electrons in the emitter and collector.

Let us apply a voltage across the base-emitter junction such that it is *forward biased*. Also, let us apply a voltage across the collector-base junction such that it is *reverse biased* (see Fig. 12-6). The base-emitter voltage V_{be} only needs to be large enough to allow current to flow across the forward biased junction (from the base into the emitter). The collector-base voltage V_{cb}, on the other hand, is usually much larger. Since the collector-base junction is *reverse* biased, the size of V_{cb} does not appreciably affect the size of the current across it. The energy diagram for this biased transistor is shown in Fig. 12-7.

With the base-emitter junction forward biased, current flows across it. Electrons flow from the emitter to the base, and holes flow from the base to the emitter. Let us first see what happens to the electrons (see Fig. 12-8). When the electrons enter the base, they begin to recombine with the holes there. However, any of the electrons from the emitter which can reach the collector-base junction without recombining with holes will fall "downhill" across the junction into the CB of the collector. In fact, if the base is narrow enough, almost *all* of the electrons will get across the base into the collector without recombining with holes.

This is one of the essential features of a good transistor. We can "turn on" the current between the collector and emitter by simply forward biasing the base-emitter junction. But we do not need to put as much current into the base-emitter junction from the base as we would when we normally forward-bias such a junction in a diode. When we forward-bias the junction in a transistor, almost all the current goes between the collector and emitter. However, no matter how well we design the transistor, *some* electrons passing through the base will recombine with holes there. This depletes the holes in the base. The only way

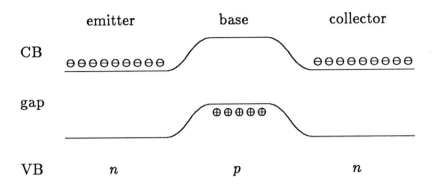

Fig. 12-5. Unbiased n-p-n transistor.

to replenish the holes in the base is for electrons in the VB to exit via the wire connected to the base. We want to make this base current as small as possible. Thus, we make the base very narrow so that most of the electrons get through without recombining (1 μm or less in Si transistors).

Something else also depletes the holes in the base, though. Since the base-emitter junction is forward biased, there is a "hole current" as well as an "electron current" across the junction. Holes in the VB of the base flow into the VB of the emitter. To make this hole current small, we dope the p-type

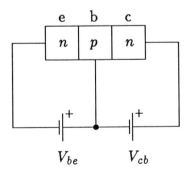

Fig. 12-6. Voltage applied to an n-p-n transistor. The base-emitter junction is forward biased, and the collector-base junction is reverse biased.

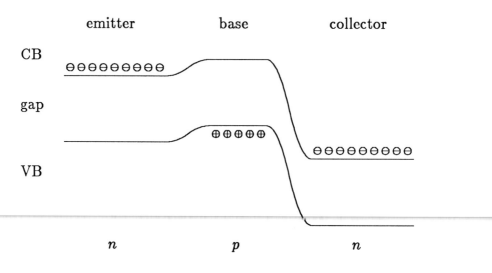

Fig. 12-7. Biased *n-p-n* transistor.

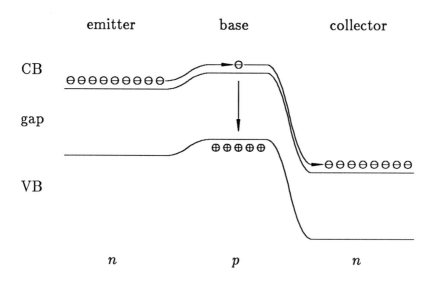

Fig. 12-8. Flow of electrons across an *n-p-n* transistor.

semiconductor in the base much more lightly than the n-type semiconductor in the emitter. This results in a much smaller density of holes in the base than the density of electrons in the emitter. Consequently, when the base-emitter junction is forward biased, the hole current across the junction is much less than the electron current, and the corresponding base current is much less than the collector current. In practice, the emitter is doped about 100 times heavier than the base.

A bipolar junction transistor is thus a current amplifier. A very small amount of current into the base causes a very large amount of current to flow into the collector. It is common for the current gain of a transistor to be as large as 100 or more. This means that for every hole which we must replenish in the base, 100 electrons get across the base into the collector.

12-3 Field-Effect Transistor

Another type of transistor is called a field-effect transistor (FET). One form of the device, known as the junction-gate field-effect transistor (JFET), consists of three layers in a semiconductor doped either n-p-n (called a p-channel JFET) or p-n-p (called an n-channel JFET). The middle layer is connected to a wire at each end, called the **source** and **drain**. The outer layers are connected to a common wire, called the **gate** (see Fig. 12-9).

We will consider the n-channel JFET here. If we apply a voltage from the drain to the source, the n-type layer will conduct a current. The magnitude of the current depends on the conductivity of the n-type semiconductor as well as the width of the channel. The conductivity depends on the doping level of the n-type layer in the semiconductor and is determined at the time of manufacturing the device. But the width of the channel can actually be controlled with the voltage we apply to the *gate*.

There is a depletion layer at each of the two p-n junctions. In the depletion layer, no free electrons or holes are present. The width of the depletion layer depends on the voltage V_a which we apply across the junctions [see Eq. (11-34)]. If we reverse-bias these junctions ($V_a < 0$), these depletion layers

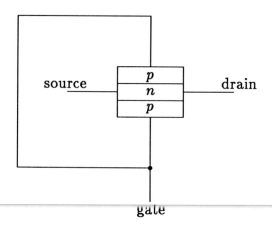

Fig. 12-9. An n-channel JFET.

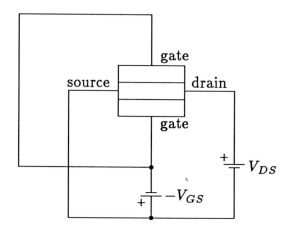

Fig. 12-10. Voltages applied to bias an n-channel JFET. The p-n junctions are reverse biased.

become wider. We do this by applying a negative voltage to the gate with respect to the source (see Fig. 12-10).

To see what this does to the drain-source current along the n-channel, consider the energy diagram across the junction. In Fig. 12-11a, we show the JFET with unbiased junctions ($V_a = 0$). In Fig. 12-11b, we show it with reverse biased junctions. The depletion layer is wider, and the width of the n-channel is *narrower*. There is a smaller number of free electrons in the n-channel, and the drain-source current is thus smaller. If we reverse-bias the junctions with even a greater voltage, the depletion layer becomes wider yet, and the drain-source current is even smaller. If we apply a large enough voltage on the gate, the two depletion layers will meet. The channel will contain *no* free electrons, and *no* current will flow. Since the p-n junctions are reverse biased during the operation of a JFET, almost no current crosses the junctions, and the gate requires very little current to operate.

To understand how the JFET can be used as an amplifier, we need to consider one more effect. When current is flowing down the channel, there is a voltage drop from the drain to the source. The channel is at a higher voltage near the drain than near the source. Thus, the p-n junctions are *more* reverse biased near the drain. This causes a *wider* depletion layer and a *narrower* channel there (see Fig. 12-12). As we increase the current down the channel, the voltage drop increases, "pinching" off the channel even more. Under these conditions, small changes in the gate voltage dramatically change the width of the channel in the "pinched" region and hence the current flowing through the channel. We thus have a device which can amplify a small voltage change at the gate into a large current change at the drain.

Another type of FET is the metal-oxide-semiconductor field-effect transistor (MOSFET). The MOSFET (enhancement mode) is very similar to the JFET except that the two outer layers of semiconductor are replaced by two layers of insulating material (like silicon dioxide), as shown in Fig. 12-13. On top of the insulators, we deposit a layer of metal which

(a) Unbiased junctions

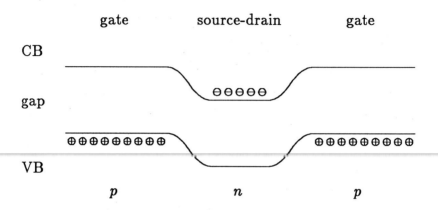

(b) Reverse biased junctions

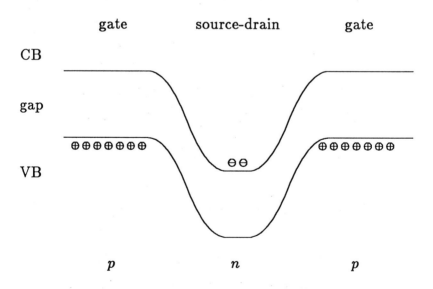

Fig. 12-11. Energy levels in an n-channel JFET.

CHAPTER 12 SEMICONDUCTOR DEVICES 253

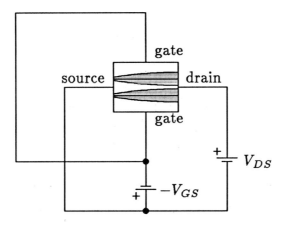

Fig. 12-12. "Pinching" due to voltage drop along the channel. The shaded regions are the depletion layers.

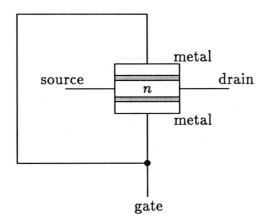

Fig. 12-13. A MOSFET. The shaded region is an insulator.

forms the gate. In the MOSFET, a negative voltage on the gate causes an electric field in the n-type semiconductor which repels electrons away from the edges of the channel. This effectively forms a depletion region there. The operation of the MOSFET is thus essentially the same as the JFET.

12-4 Metal-Semiconductor Junctions

Let us next consider a junction between a *metal* and a semiconductor. When two materials come into contact, the energies at their surfaces are initially equal. The energy difference between the Fermi level and the surface is called the **work function** Φ. If an electron is at the Fermi level, the work function is the energy required to completely remove the electron from the material.

Consider a metal with work function Φ_M in contact with an n-type semiconductor with work function Φ_S. Let $\Phi_M < \Phi_S$. When initial contact is made, the energies at their surfaces are equal (see Fig. 12-14a). Since $\Phi_M < \Phi_S$, the Fermi level of the metal is higher than the Fermi level of the semiconductor. Electrons will flow from the metal into the CB of the semiconductor. Similar to the p-n junction, this flow of electrons will cause the semiconductor to become negatively charged, producing an electric field at the junction and repelling any further flow of electrons. When the junction is at equilibrium, there will be a contact potential energy $\Phi_S - \Phi_M$, and the Fermi level will be the same on each side of the junction (see Fig. 12-14b). If we apply a voltage to this junction, electrons will freely flow between the metal and the CB of the semiconductor. Current can flow in either direction (unlike a p-n junction). This is called an **ohmic contact** (it does not rectify).

Something very different happens if $\Phi_M > \Phi_S$. In this case the Fermi level of the semiconductor is higher than the Fermi level of the metal (See Fig. 12-15a). Electrons will flow from the CB of the semiconductor to the metal, and the metal will become negatively charged (See Fig. 12-15b). Now, in order for electrons to pass between the CB of the semiconductor and the metal, they must go over a potential barrier. This

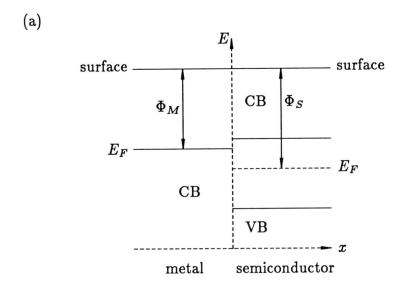

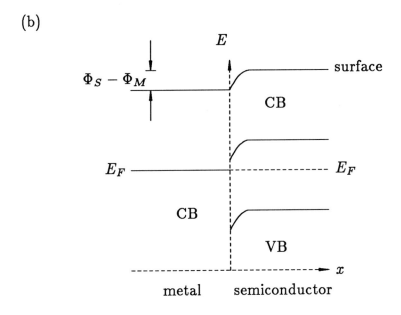

Fig. 12-14. Junction between a metal and an n-type semiconductor for the case where $\Phi_M < \Phi_S$. (a) At initial contact, energies at surfaces are equal. (b) At equilibrium, Fermi levels are equal.

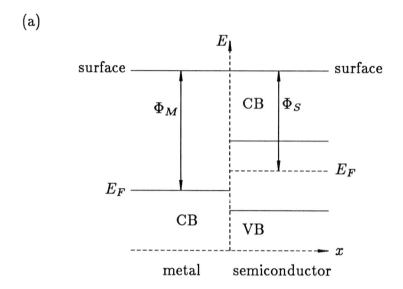

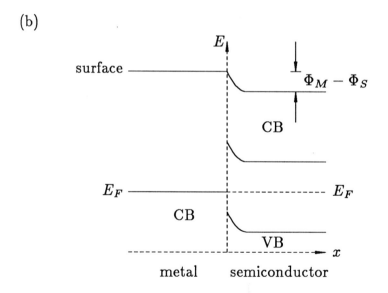

Fig. 12-15. Junction between a metal and an n-type semiconductor for the case where $\Phi_M > \Phi_S$. (a) At initial contact, energies at surfaces are equal. (b) At equilibrium, Fermi levels are equal.

(a)

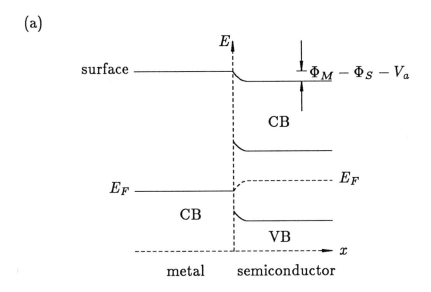

(b)

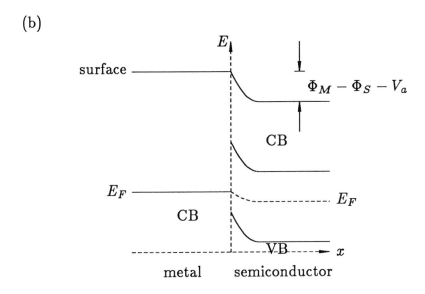

Fig. 12-16. Junction between a metal and an n-type semiconductor for the case where $\Phi_M > \Phi_S$. (a) Forward biased and (b) reverse biased junction.

situation is very much like that encountered in a *p-n* junction. If we apply a positive voltage V_a to the metal side of the junction, the barrier presented to electrons flowing from the semiconductor into the metal is decreased, and the junction is "forward biased" (see Fig. 12-16a). If we apply a positive voltage to the semiconductor side of the junction (V_a is negative), the barrier presented to electrons flowing from the metal into the semiconductor is unchanged, and very little current flows. The junction is "reverse biased" (see Fig. 12-16b). This metal-semiconductor junction is a **rectifying contact**. Such a contact is used in a device called a **Schottky-barrier diode**. An aluminum-silicon (Al-Si) junction forms a rectifying contact. The work function of Al is $\Phi_M = 4.25$ eV. The work function of *n*-type Si is $\Phi_S \cong 3.5$ eV. (Actually, the Al-Si junction can be made ohmic if the Si is very heavily doped. This technique is used for making ohmic contacts in integrated circuits.)

Problem 12-2. Show that the junction between a metal and a *p*-type semiconductor is ohmic if $\Phi_M > \Phi_S$ and rectifying if $\Phi_M < \Phi_S$. Draw an energy diagram for each of these types of junctions.

The work function for *p*-type Si is slightly less than that for Al. Consequently, as can be seen in the problem above, the junction between these two materials is ohmic.

12-5 Optical Absorption

Let us next consider some semiconductor optical devices. Light is an electromagnetic wave which we can see with our eyes. In Fig. 12-17 we show the wavelengths and their corresponding colors to which our eyes are sensitive. Electromagnetic waves with wavelengths just shorter than visible light are called **ultraviolet**. Those with wavelengths just longer than visible light are called **infrared**.

For the effects we will discuss here, we will use the "photon model" of light (which we discussed in Chapter 5). A light wave

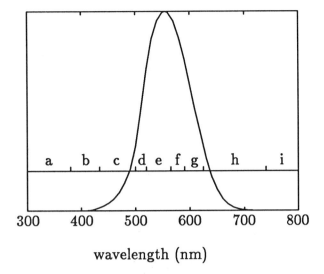

Fig. 12-17. Sensitivity of our eyes to light. The regions indicated are (a) ultraviolet, (b) violet, (c) blue, (d) cyan, (e) green, (f) yellow, (g) orange, (h) red, (i) infrared.

of frequency ω and wave vector **k** consists of photons, each with energy $\hbar\omega$ and momentum $\hbar\mathbf{k}$. We showed in Chapter 5 how a photon can collide inelastically with an electron, giving some of its energy and momentum to the electron (Compton effect). It is also possible for an electron to completely *absorb* a photon. After the absorption, the electron has additional energy $\hbar\omega$ and additional momentum $\hbar\mathbf{k}$. If the electron is in a crystal, the extra momentum $\hbar\mathbf{k}$ causes the electron to move in k-space by an amount **k**, the wave vector of the photon. Actually, the momentum of a photon in visible light is negligible and does not even need to be considered here.

Problem 12-3. If an electron absorbs a photon of wavelength 550 nm, how far does the electron move in k-space? Answer: 1.14×10^{-3} Å^{-1}.

Let us consider an electron in the VB of a semiconductor. If the electron absorbs a photon, it may have enough energy to jump up to the CB. To do this, the energy of the photon must be at least as large as the gap energy E_g. In fact, the electron cannot absorb a photon with less energy because there are no states in the gap for the electron to occupy with that energy.

Problem 12-4. What is the maximum wavelength of light that will excite electrons from the VB into the CB in silicon (Si)? (See Appendix 5.) Answer: 1110 nm.

Photons with energy greater than E_g can also be absorbed by the electrons. However, if the energy of the photon is too great, it may cause the electron to have enough energy to completely leave the crystal. If we want to excite electrons into the CB, then there is usually a useful range of energies of photons which will accomplish this. For silicon, this range is from 1.12 eV ($= E_g$) up to about 3 eV. This corresponds to light with wavelengths from 1110 nm (infrared) down to about 400 nm, which includes the entire range of visible light.

12-6 Photosensitive Devices

Let us see what happens to the conductivity of a semiconductor when we shine light on it. Some of the photons are absorbed by electrons in the VB, giving them enough energy to jump up into the CB, leaving holes behind in the VB. With extra electrons in the CB and extra holes in the VB, the conductivity in increased. Under suitable circumstances, this increase can be substantial. Such a device is called a **photoconductive cell**. One common use is a light meter for cameras. A common semiconductor used for this application is cadmium sulfide (CdS) since its highest sensitivity is in the visible-light range.

Problem 12-5. At what wavelength does a photon have the required energy E_g to excite an electron into the CB in cadmium sulfide (CdS)? (See Appendix 5.) Answer: 513 nm.

Light may also affect the operation of a diode or transistor. Consider a diode which is reverse biased. Very little current flows. If we expose the p-n junction to light, we excite electrons into the CB, leaving holes in the VB. The electrons are swept into the n-side of the junction, and the holes into the p-side, thus giving rise to a current. This device is called a **photodiode**.

A similar effect is observed in bipolar junction transistors. If we apply a voltage across the collector and emitter of a transistor but leave the base open-circuited (not connected to any wire, as shown in Fig. 12-18), we will, of course, get almost no current across the transistor. The collector-base junction is a reverse biased p-n junction.

If we shine light on this collector-base junction, we create some electron-hole pairs, just like in the photodiode. The electrons are swept into the collector and create a small current, as in the photodiode. The holes are swept into the base. These excess holes in the base attract electrons from the emitter. The

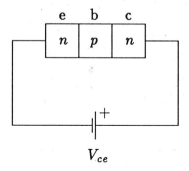

Fig. 12-18. Biased phototransistor. The base (b) is open-circuited.

electrons come into the base trying to balance the extra positive charge there. But, as we already discussed for transistors, most electrons that come into the base from the emitter do not stay in the base long enough to recombine with holes but are swept into the collector. Thus, a large number of electrons have to be brought through the base before all the extra holes there are neutralized. Most of those electrons are swept into the collector where they leave the transistor as current. Thus, a single electron-hole pair created in the collector-base junction may cause 100 or more electrons to come across the junction. The current we can obtain in such a transistor (called a **phototransistor**) is much larger than that in a photodiode. The effect of the light is multiplied by the gain of the transistor.

Suppose next that we take a photodiode with *no* voltage applied to it and shine light on it. Electron-hole pairs will be created at the p-n junction, and the electric field there in the depletion layer will force the electrons into the n-side of the junction and the holes into the p-side. Thus, a voltage develops between the two ends of the diode. If we connect the diode to a load, current will flow. Continued electron-hole pair creation at the junction will maintain a voltage and current across the load, as long as we shine light on it. The photodiode acts like a battery. When used in this way, such a device is called a **solar cell**. They are usually made of silicon. Solar cells have very important applications in the generation of electrical power from sunlight, particularly in satellites and space vehicles.

12-7 Light-Emitting Diode

Semiconductors can also be used to *produce* light as well as detect it. When an electron goes from the CB to the VB and recombines with a hole, it gives up energy. Often, this energy is given up in the form of a photon. The electron emits a photon with an energy approximately equal to the gap energy E_g. The color of light produced depends on the gap energy.

To produce enough light to be seen, we need a situation where there are simultaneously a large number of electrons in the CB and large number of holes in the VB so that there will be a large number of recombinations, producing many photons. Ordinarily, this situation is not possible, since a large number of electrons in the CB usually requires a *small* number of holes in the VB and vice versa.

However, consider a p-n junction where the semiconductor crystal on both sides of the the junction has been doped so heavily that the Fermi level is in the CB on the n-side and in the VB on the p-side (see Fig. 12-19a). These semiconductors with the Fermi level *inside* a band have properties similar to metals. Such semiconductors are said to be **degenerate**. The density of electrons in the CB on the n-side is very high. Likewise, the density of holes in the VB on the p-side is also very high.

If this junction is forward biased, there will be a region where high densities of electrons in the CB and high densities of holes in the VB exist simultaneously (see Fig. 12-19b). This of course is not an equilibrium condition. Electrons and holes recombine vigorously at the junction, producing large numbers of photons. This device is called a **light-emitting diode** (LED) and is commonly used in calculators and other electronic instruments where a low-power light source is needed.

Choosing a suitable semiconductor for an LED involves more than just finding one with the desired gap energy. There is also another important consideration. In Chapter 10, we saw that the minimum of the CB in silicon was *not* at $\mathbf{k} = 0$, but at $k = 0.99$ Å^{-1} in the [100] direction (see Fig. 10-11). The electrons in the CB are at **k**-states near this minimum. The holes in the VB, however, are at $\mathbf{k} = 0$, since that is where the maximum in the VB is. In order for an electron in the CB to fall into the VB in Si, the electron must change its wave vector **k** as well as its energy. Thus, the *crystal momentum* $\hbar\mathbf{k}$ of the electron must change.

Remember that $\hbar\mathbf{k}$ is not the *true* momentum of the electron. The true momentum of the electron is given by [see Eq. (9-1)]

$$p = mv = \frac{m}{\hbar}\frac{dE}{dk}. \qquad (12\text{-}1)$$

Since the derivative dE/dk is very small near a minimum or

264　　　　　　　　　　　　　CHAPTER 12　SEMICONDUCTOR DEVICES

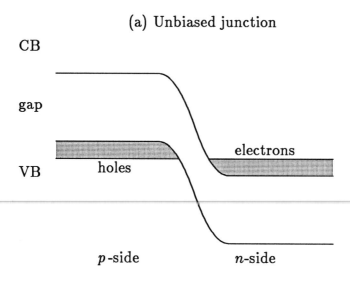

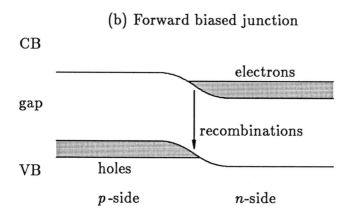

Fig. 12-19. The *p-n* junction between two degenerate semiconductors.

maximum in E, we see that the electrons both at the bottom of the CB and at the top of the VB have very small momentum. Thus, the *true* momentum of an electron does not change very much when it falls from the CB to the VB.

However, the *crystal* momentum $\hbar\mathbf{k}$ of the electron does change by a great amount when the electron falls from the CB to the VB in Si. Quantum mechanics requires that crystal momentum be *conserved*. The electron can most easily meet this requirement by emitting a *phonon* (a lattice vibration) of wave vector $\mathbf{k}$. Thus, when an electron falls from the CB to the VB in Si, it simultaneously emits a *photon* with energy E_g and a *phonon* with wave vector $\mathbf{k}$ of magnitude 0.99 Å^{-1} and direction along [100]. (The energy of the phonon is much smaller than E_g and can be neglected here.)

In gallium arsenide (GaAs), on the other hand, the bottom of the CB and the top of the VB are *both* at $\mathbf{k} = 0$. An electron falling from the CB to the VB does *not* change its wave vector $\mathbf{k}$ and only needs to emit a photon. Because of this, electrons in GaAs recombine with holes much more readily than they do in Si. GaAs is an example of a **direct-gap** semiconductor. The bottom of the CB and the top of the VB are at the same wave vector $\mathbf{k}$. Si is an example of an **indirect-gap** semiconductor. The bottom of the CB and the top of the VB are *not* at the same wave vector $\mathbf{k}$. To produce enough light to be seen in an LED, we need to use a direct-gap semiconductor.

If we want an LED to produce *visible* light, the semiconductor must have a suitable gap energy E_g.

Problem 12-6. What wavelength light would LED's made of gallium arsenide (GaAs) produce? Repeat for gallium phosphide (GaP). (See Appendix 5.) Answer: 867 nm, 549 nm.

As we can see in the above problem, an LED made of GaAs would produce light of wavelength 868 nm which is infrared and not visible. GaP looks like it might be a good choice since it should produce light of wavelength 550 nm which is green-yellow light (see Fig. 12-17). However, GaP is an *indirect-gap* semiconductor and would not produce enough light to be seen. The semiconductor most commonly used in an LED is an alloy of GaAs and GaP called gallium arsenide phosphide, $GaAs_{0.6}P_{0.4}$. It is direct-gap and produces light with a wavelength of about 650 nm which is red (see Fig. 12-17).

Problem 12-7. An LED made of gallium arsenide phosphide produces light of wavelength 650 nm. What is the width of the energy gap in this crystal? Answer: 1.91 eV.

12-8 Lasers

As a final topic in this chapter, we will discuss lasers. Consider two electron states, 1 and 2, with energies, E_1 and E_2, respectively. We label these states such that $E_2 > E_1$. Electrons prefer to be in state 1 which has the lower energy. Any electron which happens to be in state 2 can spontaneously fall into state 1, emitting a photon of energy $\hbar\omega = E_2 - E_1$. This process is called **spontaneous emission** (see Fig. 12-20). If an electron is in state 1, it is possible that it may absorb a photon of energy $\hbar\omega = E_2 - E_1$ (if any are present) and jump up to state 2. This process is called **absorption** (see Fig. 12-21).

A third process may also occur if photons of energy $\hbar\omega = E_2 - E_1$ are present, and an electron is in *state 2*. The photon can stimulate the electron to fall down into state 1. As a result, an additional photon of energy $\hbar\omega = E_2 - E_1$ is emitted (see Fig. 12-22). Both photons come out in phase with the incident photon. This process is called **stimulated emission**.

Consider a solid with many electrons in states of energy E_2. If a single photon with the correct energy $\hbar\omega = E_2 - E_1$ is present, it can stimulate one of the electrons to fall into a state of energy E_1. That process creates another photon so that there are now two photons of energy $\hbar\omega = E_2 - E_1$. Each of these may stimulate another electron to a lower state

CHAPTER 12 SEMICONDUCTOR DEVICES   267

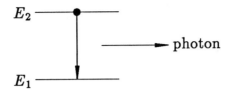

Fig. 12-20. Spontaneous emission.

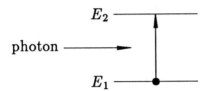

Fig. 12-21. Absorption.

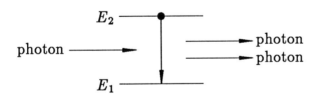

Fig. 12-22. Stimulated emission.

so that there now four photons. Each stimulated emission creates another photon, causing their number to quickly multiply, like a chain reaction. Soon there are a large number of photons present, each with energy $\hbar\omega = E_2 - E_1$. Also, each photon is in phase with the original photon, and therefore they are all in phase with each other. This system amplifies light. We start with one photon and end up with a large number. The word "laser" stands for **L**ight **A**mplification by **S**timulated **E**mission of **R**adiation.

In order for lasing to take place, there must be more electrons in states of energy E_2 than in states of energy E_1. Electrons in states of energy E_1 can absorb photons. If there are more electrons in these states than in states of energy E_2, then a photon would more likely be absorbed than stimulate emission. Photons could not be amplified. Somehow we must create a situation where more electrons are in states of energy E_2 than in states of energy E_1. This situation is called a **population inversion**.

Another necessary condition for lasing is that spontaneous emission must be very slow. In many cases, electrons can drop from a high-energy to a low-energy state very quickly by spontaneous emission. Often this happens in about a nanosecond. We could not use these states for a laser since the inverted population would be lost before any stimulated emission could take place. We need high-energy states where transitions to lower-energy states proceed very slowly, of the order of milliseconds. Such states are called **metastable**.

The first laser was built by T. Maiman in 1960. He used ruby, which is Al_2O_3 with a few chromium (Cr^{+3}) ions added (between 1 part in 1000 and 1 part in 10,000). The electron states associated with each Cr^{+3} are shown in Fig. 12-23. The lowest state (energy E_1) is normally occupied by an electron. There is an unoccupied state of energy E_2 as well as two bands of unoccupied states, labeled E_3 in the figure. We create population inversion of E_1 and E_2 using the following method. We first excite electrons from E_1 to the bands at E_3 by photon absorption. Usually a xenon flash lamp is used. The lamp is

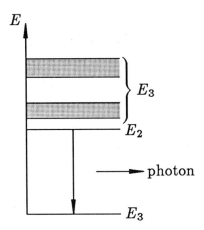

Fig. 12-23. Electron states of Cr^{+3} ions in ruby.

flashed in the vicinity of the ruby. Photons from the flash are absorbed by electrons in E_1, exciting them to the energy levels in E_3. As soon as an electron is excited to one of these levels, it quickly (within 1 ns) drops down to the state E_2 by emitting a *phonon*. The state at E_2 is metastable, so its occupancy increases as the flash continues to excite electrons out of the state E_1. When population inversion has been finally achieved (more electrons in E_2 than in E_1), stimulated emission causes the ruby to lase and emit a pulse of light.

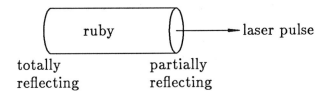

Fig. 12-24. Ruby laser.

Problem 12-8. In Fig. 12-23, $E_2 - E_1 = 1.786$ eV. Find the wavelength of the light given off by a ruby laser. Answer: 694 nm.

In practice, the ruby crystal has the shape of a cylinder (see Fig. 12-24). The ends are polished and silver-coated so that one end is totally reflecting and the other end is partially reflecting, allowing about 5% of the light to leave the crystal. Thus, a photon usually bounces back and forth many times between the two ends of the crystal before it finally leaves. This allows each photon to stimulate a larger number of emissions and increases the power output of the laser.

Problem 12-9. Suppose that total population inversion could be achieved in a ruby laser. If half of the electrons in E_2 could then drop to E_1 in 30 ns, what would be the average power of the resulting laser pulse over this time interval? Assume that the ruby crystal is a cylinder 5.00 cm long and 1.00 cm in diameter and that one Al in every 2000 has been replaced by a Cr. The density of Al_2O_3 is 3.7 g/cm^3. Use $E_2 - E_1 = 1.786$ eV in Fig. 12-23. Answer: 410 MW.

The **diode laser** (or injection laser) also uses the principle of population inversion. We see in Fig. 12-19b that a population inversion is achieved when an LED is forward biased. In

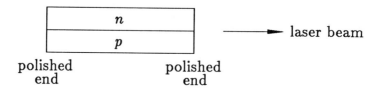

Fig. 12-25. Diode laser.

this case, E_1 is at the top of the VB, and E_2 is at the bottom of the CB. The LED can be made into a laser by polishing two ends so that 30% of the emitted photons are reflected back into the crystal (see Fig. 12-25). As the number of photons at the p-n junction increases, the probability of a stimulated emission becomes greater. When stimulated emission dominates over the usual spontaneous emission of an LED, we have a laser. Diode lasers can be made very small (of the order of a millimeter in dimension). These devices are very useful for optical communications using optical fibers.

CHAPTER 13

SUPERCONDUCTIVITY

13-1 Introduction

In 1908, K. Onnes at the University of Leiden in Holland succeeded in liquefying helium. Liquid helium boils at 4.2 K. The availability of liquid helium made it possible to study phenomena at a much lower temperature than previously possible. Three years later, in 1911, he was measuring the resistivity of mercury metal as a function of temperature and discovered that at about 4 K, the resistivity suddenly dropped to an unmeasurably small value (see Fig. 13-1). We call this phenomenon **superconductivity**, and we now know that the resistivity of a superconductor really is zero. The transition temperature to the superconducting state is called the **critical temperature** T_c. Other metals were soon found to also exhibit superconductivity, and today more than twenty elements and thousands of alloys are known superconductors. Selected

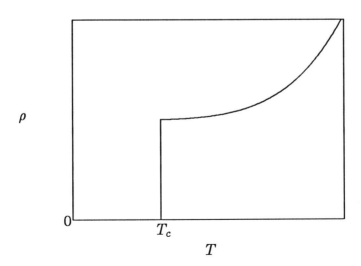

Fig. 13-1. The resistivity ρ of a sample of mercury (Hg) as a function of temperature T.

examples of these are listed in Appendix 7. Note that only non-magnetic metals can exhibit superconductivity.

13-2 Trapped Magnetic Flux and Persistent Currents

Let us discuss some of the unusual properties of superconductors. Consider a loop of wire made of a metal which exhibits superconductivity. At a temperature $T > T_c$, the metal is in the non-superconducting state, called the **normal state**. We place this wire loop in a magnetic field **B** and then lower the temperature of the wire below T_c so that it becomes superconducting. You may recall that Faraday's law of induction is

$$\oint \mathcal{E} \cdot d\mathbf{l} = -d\Phi_B/dt. \qquad (13\text{-}1)$$

This equation means that the line integral of the electric field $\mathcal{E}$ around any closed loop will be equal to the negative rate of change in the magnetic flux Φ_B through the loop. Since the resistivity $\rho = 0$ inside the superconductor, then $\mathcal{E} = 0$ there also (otherwise, the current density $\mathbf{J} = \mathcal{E}/\rho$ would be infinite). If we consider a line integral along a path inside the wire around the loop, we see that since $\mathcal{E} = 0$ everywhere along the path, the integral is zero, and $d\Phi_B/dt = 0$. This means that the magnetic flux through the wire loop cannot change. The flux

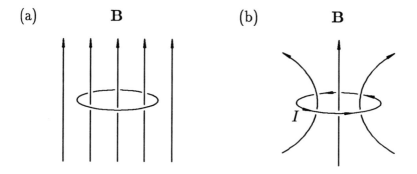

Fig. 13-2. (a) A loop of superconducting wire in an external magnetic field **B**. (b) When the external field is turned off, a current I appears in the wire, generating its own field **B** through the loop.

CHAPTER 13 SUPERCONDUCTIVITY   275

is "trapped." If we now turn off the magnetic field, how does the superconducting wire prevent Φ_B from going to zero? A current spontaneously appears in the wire, generating its own field **B** in the loop such that the flux Φ_B is the same as before we turned off the field (see Fig. 13-2). If the resistance of the wire is truly zero, this current will persist forever. We can detect any decay in this current by measuring the magnetic field it generates. Such currents have been observed to persist for up to *three years* without any detectable decay. Experimentally, it has been shown that the conductivity σ of a superconductor is greater than $10^{26}/\Omega\cdot\text{m}$!

13-3 Meissner Effect

A superconductor is actually more than just a perfect conductor. It is also a perfect **diamagnet**. This means that **B** = 0 as well as **ε** = 0 everywhere inside the superconductor. This property is not simply a consequence of $\rho = 0$. Since **ε** = 0 everywhere inside the superconductor, we see from Faraday's law in Eq. (13-1) that the magnetic flux Φ_B through *any* closed path inside the superconductor cannot change with time ($d\Phi_B/dt = 0$). The only way this can be true is that the magnetic field **B** inside the superconductor be constant ($d\mathbf{B}/dt = 0$). Magnetic fields can penetrate a metal when it is in its normal state. If we lower the temperature below T_c while the metal is sitting in an external magnetic field, Faraday's law would suggest that the field inside the superconductor would remain constant ($d\mathbf{B}/dt = 0$) and be trapped there. This does not happen. When the metal becomes superconducting, the field is actually expelled so that **B** = 0 everywhere inside. This is called the **Meissner effect**, discovered by W. Meissner and R. Ochsenfeld in 1933.

A conductor expels *electric* fields by moving electric *charges* to its surface. These surface charges generate an electric field which exactly cancels the externally applied electric field everywhere inside the conductor. Similarly, a superconductor expels *magnetic* fields by setting up electric *currents* at its surface. These surface currents generate a magnetic field which exactly cancels the externally applied magnetic field everywhere inside the superconductor. When the temperature is lowered below T_c and the metal becomes superconducting, these surface currents spontaneously appear. If we change the external field, these surface currents also change, maintaining $B = 0$ everywhere inside the superconductor.

13-4 Penetration of Magnetic Fields

The surface currents cannot flow in an infinitely thin layer on the surface. They must penetrate the superconductor to some extent. Consequently, an external magnetic field also penetrates the surface of a superconductor. It is found that the field near the surface of a superconductor is given by

$$\mathbf{B} = \mathbf{B}_0 \exp(-x/\lambda), \qquad (13\text{-}2)$$

where $\mathbf{B}_0$ is the magnetic field at the surface, x is the distance from the surface, and λ is called the **penetration depth** (see Fig. 13-3). Also, the direction of $\mathbf{B}_0$ must be parallel to the surface. As we can see, the magnetic field goes to zero at depths a few times λ below the surface. The value of λ is typically less than 1000 Å.

The penetration depth λ also depends on temperature. An empirical expression for λ is given by

$$\lambda = \frac{\lambda_0}{\sqrt{1 - (T/T_c)^4}}. \qquad (13\text{-}3)$$

This function is plotted in Fig. 13-4. We see that λ becomes infinite as T approaches T_c. The magnetic field penetrates deep into the superconductor when the temperature is near T_c. When $\mathbf{B}$ penetrates the entire sample ($\lambda \to \infty$) then the sample loses its superconductivity.

Problem 13-1. For lead (Pb), we find that $T_c = 7.193$ K and $\lambda_0 = 390$ Å. Find the penetration depth λ at $T = 0$, 3 K, and 7.100 K. Answer: 390 Å, 396 Å, 1730 Å.

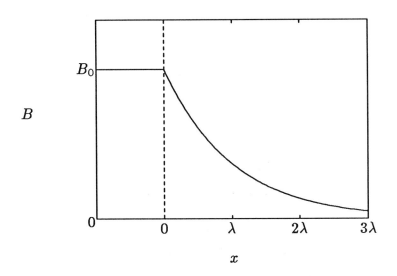

Fig. 13-3. The magnetic field B inside a superconductor as a function of the distance x from the surface. The field outside the superconductor ($x < 0$) is equal to B_0.

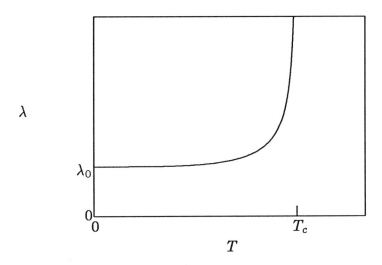

Fig. 13-4. The penetration depth λ as a function of temperature T.

13-5 Critical Fields

Soon after Onnes discovered superconductivity in mercury (Hg), a number of other metals, including lead (Pb), were also found to exhibit this phenomenon. The idea was conceived that a solenoid made of superconducting wire might be capable of generating very large magnetic fields if the wire could carry very large currents. (We call such a device a **superconducting magnet.**) It was soon discovered, though, that magnetic fields could destroy superconductivity. When the field B at the surface of the superconductor exceeds some **critical field** B_c, the metal becomes normal. It is found that the critical field B_c depends on the temperature in the following way:

$$B_c(T) = B_c(0) \left[1 - (T/T_c)^2\right]. \tag{13-4}$$

This function is plotted in Fig. (13-5). The largest value of $B_c(T)$ occurs at $T = 0$. If the field exceeds $B_c(0)$, the metal will not go superconducting at *any* temperature, not even at $T = 0$. [See Appendix 7 for values of $B_c(0)$.]

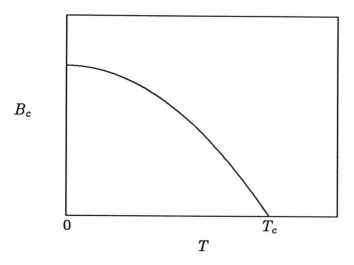

Fig. 13-5. The critical field B_c as a function of temperature T.

Problem 13-2. Find the critical field in lead (Pb) at T=4.2 K. Answer: 0.053 T.

We see in the above problem that in liquid helium (4.2 K), lead will be superconducting only in fields less than 0.053 T. This field is not very large and can be easily obtained in electromagnets using normal wire. Superconducting magnets did not seem very practical under these conditions.

Notice that the current which a superconducting wire can carry is also limited. You may recall that the field generated a distance R from a wire carrying a current I is given by

$$B = \mu_0 I / 2\pi R. \tag{13-5}$$

At a certain critical current I_c, the magnetic field B at the surface of the wire ($R = r$, the radius of the wire) will be equal to B_c. From Eq. (13-5), we see that this happens at

$$I_c = 2\pi r B_c / \mu_0. \tag{13-6}$$

The current produces a field which causes the wire to go to its normal state.

Problem 13-3. How much current can a lead (Pb) wire, 1.00 mm in diameter, carry in its superconducting state at 4.2 K? You may use the result of Problem 13-2. Answer: 130 A.

13-6 Type II Superconductors

In the 1950's, it became apparent that there were two kinds of superconductors, called type I and type II. The superconductors we have been considering thus far are type I. In type II superconductors, there are *two* kinds of critical fields, B_{c1} and B_{c2} (see Fig. 13-6). At fields below B_{c1}, the metal is entirely superconducting, just like a type I superconductor. At

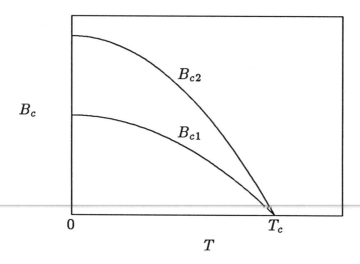

Fig. 13-6. The critical fields B_{c1} and B_{c2} as a function of temperature T in a type II superconductor.

fields above B_{c2}, the metal is normal. Between B_{c1} and B_{c2}, the metal is in a **mixed state**. "Threads" of normal metal appear throughout the superconductor, and the magnetic field can penetrate the metal along these threads. In order to keep the field out of the superconducting regions of the metal, current circulates around these threads. For this reason, these threads are called vortices, and this mixed state is called the vortex state.

As the field is increased, more of these threads of normal metal appear, until, at B_{c2}, the entire metal becomes normal. As long as there are any regions of superconducting metal present, the resistivity ρ of the sample is zero. Thus, the sample is essentially superconducting at all fields below B_{c2}. The critical field B_{c2} in a type II superconductor is usually quite large. (See Appendix 7 for values of B_{c2}.) For example, a certain alloy of niobium and titanium (75% Nb and 25% Ti by weight) has a critical field $B_{c2} = 10$ T at $T = 4.2$ K. This is a popular alloy for the construction of superconducting magnets.

Problem 13-4. Consider a solenoid of copper wire, 1.00 mm in diameter, with 100 winds per cm. (This solenoid would be wound about 10 layers thick.) The diameter of the solenoid is 10 cm and its length is 20 cm. How much current would be required to obtain a field $B = 1.00$ T in this solenoid? How much power would be dissipated due to resistive heating of the wire? Recall that the magnetic field in a solenoid is $B = \mu_0 I n$ where I is the current in the wire, n is the number of loops of wire per unit length along the solenoid, and μ_0 is the permeability constant (see Appendix 1). (Consider the 10 cm to be the *average* diameter of the windings, and *not* the *inside* diameter.) Answer: 80 A, 86 kW.

In the above problem we illustrate the technological difficulties of generating large fields using normal wire. To obtain a field of only 1 T, we would generate 86 kW of heat due to resistive heating of the wire. This much heat is very difficult to remove from the solenoid. In contrast, a solenoid of superconducting wire (a superconducting magnet) dissipates *no* heat ($\rho = 0$) and fields greater than 10 T have been generated with these magnets. Of course, an obvious disadvantage of superconducting magnets is that the solenoid must be kept very cold, usually 4.2 K in liquid helium.

13-7 BCS Theory

Let us now discuss the underlying theory of superconductivity. This theory was first formulated by J. Bardeen, L. N. Cooper, and J. R. Schrieffer in 1957. In their honor, it is commonly known as the **BCS theory**. A rigorous understanding of this theory is way beyond the scope of this book. We will discuss here only some of its general aspects.

Electrons normally repel each other due to their negative electric charge. At low temperatures, it is possible for some electrons in a metal to actually be *attracted* to each other. They form bound pairs, called **Cooper pairs**. This attractive force arises from lattice deformations near the electrons. An electron attracts the positive ions nearby which then "screen" the negative charge of the electron (see Fig. 13-7). In fact, the positive ions may overrespond so that the region around the electron actually acquires a net positive charge. Another electron passing by could be attracted to this region. In this way, two electrons could be effectively attracted to each other. Since both electrons are moving through the lattice, these deformations are quite dynamic in nature and constitute a localized lattice vibration.

In quantum mechanics, we describe this lattice vibration by a *phonon*. The attraction between the two electrons is actually due to an exchange of phonons. One electron in the Cooper pair emits a phonon, and the other electron absorbs it. By conservation of crystal momentum, we have

$$\mathbf{k}_{1f} = \mathbf{k}_{1i} - \mathbf{k}$$

and

$$\mathbf{k}_{2f} = \mathbf{k}_{2i} + \mathbf{k}, \qquad (13\text{-}7)$$

where $\mathbf{k}$ is the wave vector of the exchanged phonon, $\mathbf{k}_{1i}$ and $\mathbf{k}_{1f}$ are the initial and final wave vectors, respectively, of the electron which emits the phonon, and $\mathbf{k}_{2i}$ and $\mathbf{k}_{2f}$ are the initial and final wave vectors, respectively, of the electron which absorbs the phonon. Note that

$$\mathbf{k}_{1f} + \mathbf{k}_{2f} = \mathbf{k}_{1i} + \mathbf{k}_{2i}. \qquad (13\text{-}8)$$

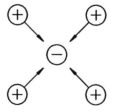

Fig. 13-7. Positive ions attracted toward an electron create a region of net positive charge.

The total crystal momentum of the Cooper pair is conserved:

$$\mathbf{k}_1 + \mathbf{k}_2 = \text{constant}. \tag{13-9}$$

A Cooper pair rapidly exchanges many phonons so that its wave function ψ is actually a superposition of many pairs of wave functions:

$$\psi = \sum_{i,j} A_{ij} \psi_1(\mathbf{k}_i) \psi_2(\mathbf{k}_j), \tag{13-10}$$

where $\psi_1(\mathbf{k}_i)$ is the Bloch function for electron #1 in the state $\mathbf{k}_i$, $\psi_2(\mathbf{k}_j)$ is the Bloch function for electron #2 in the state $\mathbf{k}_j$, and A_{ij} is the amplitude of ψ when electron #1 is in $\mathbf{k}_i$ and electron #2 is in $\mathbf{k}_j$. The summation is taken over all pairs, $\mathbf{k}_i$ and $\mathbf{k}_j$, such that $\mathbf{k}_i + \mathbf{k}_j$ is equal to some constant. It is found that the attraction in the Cooper pairs increases when there is a larger number of states $\mathbf{k}_i, \mathbf{k}_j$ which they can sample.

Let us consider a free-electron model. Let k_F be the radius of the Fermi sphere. At $T = 0$ in a normal metal, all states $k < k_F$ are occupied and all states $k > k_F$ are unoccupied. With this distribution of occupied states, no phonons can be exchanged between any pairs of electrons (keeping them still inside the Fermi surface) since they would have to end up in states already occupied by other electrons. Pauli's exclusion principle forbids this.

However, if we move a pair of electrons into states *outside* the Fermi surface ($k > k_F$), then phonon exchanges would be possible. The resulting attraction could more than offset the increase in kinetic energy ($\hbar^2 k^2/2m$) required to take the pair outside the Fermi surface. This pair then would have a *net* energy $E < E_F$ and would form a stable state.

The attraction is strongest (and the net energy lowest) when the two electrons are in states so that the maximum number of pairs of states $\mathbf{k}_i, \mathbf{k}_f$ can be sampled. Both $\mathbf{k}_i$ and $\mathbf{k}_f$ must also be near the Fermi surface so that the kinetic energy is as small as possible. For example, consider the pair

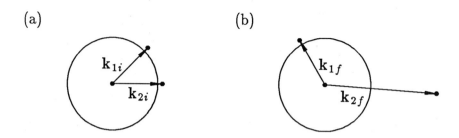

Fig. 13-8. (a) Initial and (b) final states of a pair of electrons which exchange a phonon. The circle represents the Fermi surface.

of states in Fig. 13-8a. If the electron at k_{1i} emits a phonon so that it goes to the state k_{1f} shown in Fig. 13-8b (also near the Fermi surface), then by conservation of momentum, when the electron at k_{2i} absorbs that phonon, it must end up in the state k_{2f} also shown in Fig. 13-8b. This state k_{2f} is far removed from the Fermi surface. This phonon exchange is not energetically favorable since the second electron would end up in a state of high kinetic energy. Such a phonon would not be exchanged between these electrons. It is easily seen that for the initial states shown in Fig. 13-8a, only nearby states would be sampled.

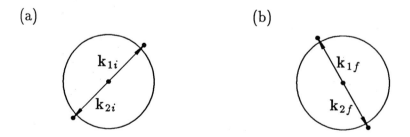

Fig. 13-9. (a) Initial and (b) final states of a pair of electrons which exchange a phonon. The circle represents the Fermi surface.

CHAPTER 13 SUPERCONDUCTIVITY   285

Problem 13-5. Show graphically that Eq. (13-8) holds for the wave vectors in Fig. 13-8. Also draw the wave vector of the exchanged phonon.

In contrast, consider two states on opposite sides of the Fermi surface ($k_{2i} = -k_{1i}$) as shown in Fig. 13-9a. The total crystal momentum $k_{1i} + k_{2i}$ is zero for this pair. If the electron at k_{1i} now emits a phonon so that it goes to the same state k_{1f} as in Fig. 13-8b, the electron at k_{2i} absorbs that phonon and goes to the state k_{2f} shown in Fig. 13-9b. (The total crystal momentum of the electrons is still zero.) The state k_{2f} is near the Fermi surface in this case, and this phonon exchange is energetically favorable. In fact *all* states around the Fermi surface can be sampled by this pair. This pair of electrons thus experiences a much greater attractive force than the pair shown in Fig. 13-8. For this reason, Cooper pairs always contain electrons with wave vectors in the opposite directions. The total crystal momentum of a Cooper pair is zero.

If one pair can lower its net energy by going to states outside the Fermi surface, others can too. As we fill up states outside the Fermi surface with Cooper pairs, though, we also limit the number of states they can sample because of the Pauli exclusion principle. A Cooper pair cannot go into any states already occupied by another Cooper pair at that instant. As a result, there is an optimum distribution of electrons into those states (see Fig. 13-10).

This distribution looks very much like the Fermi-Dirac distribution function shown in Fig. 7-6. However, in the case of a superconductor, the temperature is *zero*. Notice that since there are occupied states above k_F, there are also *unoccupied* states below k_F. Thus, Cooper pairs may also form below k_F. All the electrons near k_F form Cooper pairs with net energies $E < E_F$. The resulting density of states $g(E)$ near E_F appears like Fig. 13-11. All the Cooper pairs have energies less than $E_F - \Delta$. It requires an energy 2Δ to break up a Cooper pair.

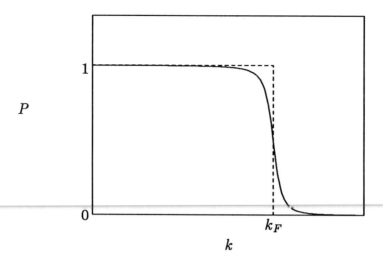

Fig. 13-10. The probability P that a state k will be occupied by an electron at $T = 0$ in a superconductor.

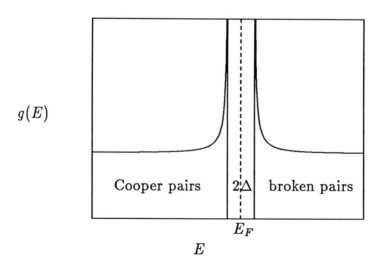

Fig. 13-11. The density of states $g(E)$ near the Fermi energy E_F in a superconductor.

The broken pair must occupy states of energy $E > E_F + \Delta$. Thus there is an *energy gap* of width 2Δ centered on E_F. This energy gap is usually of the order of 10^{-3} eV at $T = 0$. (See Appendix 7 for values of 2Δ.)

BCS theory predicts that the gap energy is proportional to the critical temperature:

$$2\Delta \cong 3.5 k_B T_c. \qquad (13\text{-}11)$$

Superconductors with large gap energies maintain superconductivity to higher temperatures.

Problem 13-6. Using Eq. 13-11 and the values of T_c in Appendix 7, calculate the gap energy for Al, Hg, and Pb. Compare with the experimental values of 2Δ in Appendix 7. Answer: 0.36×10^{-3} eV, 1.25×10^{-3} eV, 2.17×10^{-3} eV.

At temperatures above zero, some Cooper pairs are broken up due to thermal energy, and some states above $E_F + \Delta$ become occupied. Each unpaired electron occupies a state **k** which is now no longer available for sampling by the remaining Cooper pairs, due to Pauli's exclusion principle. This reduces the attractive force within these remaining Cooper pairs, causing the width of the energy gap 2Δ to decrease. As T increases further, 2Δ decreases even further, as shown in Fig. 13-12. At $T = T_c$, 2Δ goes to zero, and the superconductor goes normal. Formation of Cooper pairs is no longer energetically favorable.

When an electric field is applied to a superconductor, each Cooper pair is accelerated so that the total crystal momentum of each pair is non-zero:

$$\mathbf{k}_1 + \mathbf{k}_2 = \Delta \mathbf{k}. \qquad (13\text{-}12)$$

The value of $\Delta \mathbf{k}$ is the same for every Cooper pair. The center of mass of each pair moves with the same velocity. From the nature of the quantum mechanical state of this system, we find

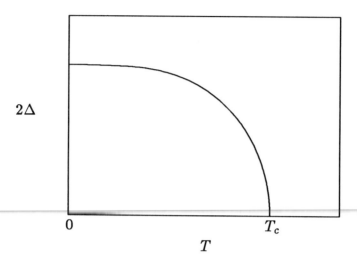

Fig. 13-12. The gap energy 2Δ of a superconductor as a function of temperature T.

that we cannot change the velocity of one Cooper pair without changing it for *all* of them. Thus, lattice imperfections (like impurities and phonons), which are effective in scattering electrons in normal metals, can have no effect on Cooper pairs. Without scattering, the current persists forever, and the resistivity is zero.

It may seem rather strange that phonons, which are responsible for resistivity in normal metals, provide the interactions which give rise to superconductivity. In fact, a strong electron-phonon interaction *favors* superconductivity. Such metals normally are *poor* conductors in their normal state. In contrast, the best conductors (like copper and silver) do not exhibit superconductivity at all.

13-8 Isotope Effect

Let us consider direct experimental evidence of the role of phonons in superconductivity. Metals made of different isotopes of the same atom become superconducting at different critical temperatures T_c. Isotopes of a given atom differ only in number of neutrons in the nucleus. Such atoms are chemically identical but have different *masses*. The difference in mass affects the frequencies of the lattice vibrations. For example, heavier atoms vibrate with lower frequencies. Thus, any phenomenon (such as superconductivity) which involves phonons should be affected by the mass of the atoms present.

For example, mercury (Hg) has seven naturally occurring isotopes. Their masses are 196, 198, 199, 200, 201, 202, and 204 amu, respectively. (These isotopes are designated ^{196}Hg, ^{198}Hg, etc.) A sample of naturally occurring Hg contains some of each of these seven isotopes. The atomic mass given on the periodic table in Appendix 2 is the average mass of the atoms in such a sample (200.59 amu for the case of Hg). Various techniques have been developed for separating these isotopes so that samples can be made which are enriched in one of these isotopes. A measurement of the critical temperature T_c in these various isotopes has shown that T_c decreases with increasing atomic mass. This is called the **isotope effect**. For example, T_c is equal to 4.161 K in ^{199}Hg and 4.126 K in ^{204}Hg ($T_c = 4.153$ K in naturally occurring Hg).

13-9 Absorption of Electromagnetic Radiation

The absorption of electromagnetic radiation by superconductors provides evidence for the energy gap. In Chapter 12, we saw that semiconductors could absorb photons which had an energy greater than the gap energy. Electrons in the valence band absorb these photons and are excited across the gap into the conduction band. Similarly, superconductors absorb photons which have an energy greater than *its* gap energy 2Δ. Cooper pairs absorb these photons and are broken apart. Photons with energy less than 2Δ will not be absorbed by a superconductor. Thus, if we measure the absorption of photons as a function of their energy $\hbar\omega$, we find that the absorption greatly increases when $\hbar\omega > 2\Delta$. Not only does this effect indicate the existence of energy gaps in superconductors but also provides a method for measuring their width 2Δ.

Problem 13-7. Find the maximum wavelength of electromagnetic radiation which will be absorbed by Hg at $T = 0$. Answer: 0.75 mm.

13-10 Tunneling

Further evidence for the existence of energy gaps in superconductors is given by tunneling experiments. For example, consider the tunneling between aluminum (Al) and lead (Pb). A strip of Al is evaporated onto a glass slide (see Fig. 13-13). A thin oxide layer (Al_2O_3) is then allowed to grow on the Al. Then a strip of Pb is evaporated across the Al. When the glass slide is immersed in liquid helium ($T = 4.2$ K), the Pb is superconducting ($T_c = 7.2$ K) while the Al is not ($T_c = 1.2$ K). The oxide layer between them normally prevents current from flowing from one metal to the other. With no applied voltage, the two metals are in equilibrium, and the Fermi energy E_F in both metals must be equal (see Fig. 13-14a). If we apply a voltage V_a across the insulating layer, there is a possibility that current may flow via *tunneling* (see Chapter 6). However, if $V_a < \Delta/e$, there are no states in Pb to which electrons from Al can tunnel (see Fig. 13-14b). The Fermi energy E_F in Al is still within the energy gap in Pb. As we increase V_a above

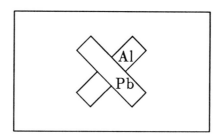

Fig. 13-13. Strip of Pb evaporated across a strip of Al for a tunneling experiment.

(a) $V_a = 0$

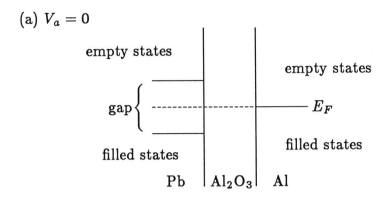

(b) $V_a < \Delta/e$

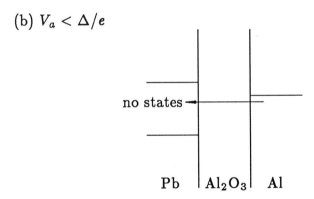

(c) $V_a > \Delta/e$

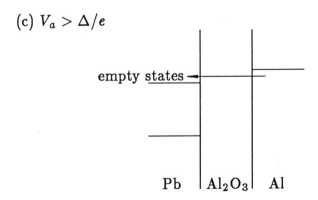

Fig. 13-14. Energy levels in superconducting Pb and normal Al when the applied voltage V_a is (a) zero, (b) between zero and Δ/e, and (c) greater than Δ/e.

Δ/e, the Fermi energy in Al is raised above the gap in Pb, and now electrons can tunnel from Al into the empty states of Pb above the gap (see Fig. 13-14c). Thus, as we turn up the voltage V_a, we see a sharp increase in the current when $V_a = \Delta/e$. In Pb, $\Delta = 1.37 \times 10^{-3}$ eV, so this increase in current occurs when $V_a = 1.37$ mV.

APPENDIX 1
SOME PHYSICAL CONSTANTS

electron charge	e	1.602×10^{-19} C
electron mass		9.11×10^{-31} kg
proton mass		1.673×10^{-27} kg
neutron mass		1.675×10^{-27} kg
Planck's constant	h	6.626×10^{-34} J·s
	$\hbar$	1.054×10^{-34} J·s
Boltzmann's constant	k_B	1.380×10^{-23} J/K
permittivity constant	ϵ_0	8.854×10^{-12} F/m
permeability constant	μ_0	1.257×10^{-6} H/m
speed of light	c	3.00×10^8 m/s
Avogadro's number		6.022×10^{23} /mol

APPENDIX 2
THE ELEMENTS

Element		Atomic Number	Atomic Mass
actinium	Ac	89	227
aluminum	Al	13	26.98154
americium	Am	95	243
antimony	Sb	51	121.75
argon	Ar	18	39.948
arsenic	As	33	74.9216
astatine	At	85	210
barium	Ba	56	137.33
berkelium	Bk	97	247
beryllium	Be	4	9.01218
bismuth	Bi	83	208.9808
boron	B	5	10.81
bromine	Br	35	79.904
cadmium	Cd	48	112.41
calcium	Ca	20	40.08
californium	Cf	98	251
carbon	C	6	12.011
cerium	Ce	58	140.12
cesium	Cs	55	132.9054
chlorine	Cl	17	35.453
chromium	Cr	24	51.996
cobalt	Co	27	58.9332
copper	Cu	29	63.546
curium	Cm	96	247
dysprosium	Dy	66	162.50

APPENDIX 2 THE ELEMENTS

Element		Atomic Number	Atomic Mass
einsteinium	Es	99	254
erbium	Er	68	167.26
europium	Eu	63	151.96
fermium	Fm	100	257
fluorine	F	9	18.998403
francium	Fr	87	223
gadolinium	Gd	64	157.25
gallium	Ga	31	69.737
germanium	Ge	32	72.59
gold	Au	79	196.9665
hafnium	Hf	72	178.49
hahnium	Ha	105	
helium	He	2	4.00260
holmium	Ho	67	164.9304
hydrogen	H	1	1.0079
indium	In	49	114.82
iodine	I	53	126.9045
iridium	Ir	77	192.22
iron	Fe	26	55.847
krypton	Kr	36	83.80
lanthanum	La	57	138.9055
lawrencium	Lr	103	
lead	Pb	82	207.2
lithium	Li	3	6.941
lutetium	Lu	71	174.967
magnesium	Mg	12	24.305
manganese	Mn	25	54.9380
mendelevium	Md	101	256

Element		Atomic Number	Atomic Mass
mercury	Hg	80	200.59
molybdenum	Mo	42	95.94
neodymium	Nd	60	144.24
neon	Ne	10	20.179
neptunium	Np	93	237.0482
nickel	Ni	28	58.71
niobium	Nb	41	92.9064
nitrogen	N	7	14.0067
nobelium	No	102	
osmium	Os	76	190.2
oxygen	O	8	15.99994
palladium	Pd	46	106.4
phosphorus	P	15	30.97376
platinum	Pt	78	195.09
plutonium	Pu	94	
polonium	Po	84	
potassium	K	19	39.0983
praseodymium	Pr	59	140.9077
promethium	Pm	61	
protactinium	Pa	91	231.0359
radium	Ra	88	226.0254
radon	Rn	86	222
rhenium	Re	75	186.2
rhodium	Rh	45	102.9055
rubidium	Rb	37	85.4678
ruthenium	Ru	44	101.07
rutherfordium	Rf	104	
samarium	Sm	62	150.4

APPENDIX 2   THE ELEMENTS

Element		Atomic Number	Atomic Mass
scandium	Sc	21	44.9559
selenium	Se	34	78.96
silicon	Si	14	28.0855
silver	Ag	47	107.868
sodium	Na	11	22.9898
strontium	Sr	38	87.62
sulfur	S	16	32.06
tantalum	Ta	73	180.9479
technetium	Tc	43	98.9062
tellurium	Te	52	127.60
terbium	Tb	65	158.9254
thallium	Tl	81	204.37
thorium	Th	90	232.0381
thulium	Tm	69	168.9342
tin	Sn	50	118.69
titanium	Ti	22	47.90
tungsten	W	74	183.85
uranium	U	92	238.029
vanadium	V	23	50.9415
xenon	Xe	54	131.30
ytterbium	Yb	70	173.04
yttrium	Y	39	88.9059
zinc	Zn	30	65.38
zirconium	Zr	40	91.22

APPENDIX 3

CRYSTAL STRUCTURES

Listed below are the crystal structures of various elements and compounds. The lattice parameters a in units of Å are also given.

Elements with a bcc lattice

Li	3.50	Ba	5.02	Mo	3.14
Na	4.30	V	3.04	W	3.15
K	5.20	Nb	3.30	Fe	2.86
Rb	5.59	Ta	3.32	Eu	4.58
Cs	6.50	Cr	2.87		

Elements with an fcc lattice

Ca	5.56	Pd	3.87	Al	4.04
Sr	6.08	Pt	3.90	Pb	4.93
Ac	5.31	Cu	3.61	Ce	5.12
Rh	3.80	Ag	4.07	Yb	5.48
Ir	3.82	Au	4.07	Th	5.08
Ni	3.52				

Elements with the diamond structure

C	3.56	Ge	5.65	Sn	6.46
Si	5.42				

Compounds with the cesium chloride structure

CsCl	4.11	TlI	4.18	CuPd	2.99
CsBr	4.28	TlSb	3.85	AgZn	3.16
CsI	4.56	TlBi	3.90	AuZn	3.15
TlCl	3.84	CuZn	2.95	AlNi	2.82
TlBr	3.97				

APPENDIX 3 CRYSTAL STRUCTURES

Compounds with the sodium chloride structure

LiF	4.02	RbI	7.32	BaTe	6.99
LiCl	5.14	CsF	6.00	MnO	4.43
LiBr	5.49	MgO	4.20	MnS	5.21
LiI	6.00	MgS	5.19	MnSe	5.45
NaF	4.61	MgSe	5.45	FeO	4.28
NaCl	5.63	CaO	4.80	CoO	4.25
NaBr	5.96	CaS	5.68	NiO	4.17
NaI	6.46	CaSe	5.91	AgF	4.92
KF	5.36	CaTe	6.34	AgCl	5.54
KCl	6.27	SrO	5.15	AgBr	5.76
KBr	6.58	SrS	6.01	CdO	4.70
KI	7.05	SrSe	6.23	SnTe	6.28
RbF	5.63	SrTe	6.65	PbS	5.93
RbCl	6.53	BaO	5.53	PbSe	6.14
RbBr	6.85	BaSe	6.59	PbTe	6.44

Compounds with the zincblende structure

BeS	4.86	ZnTe	6.09	GaP	5.44
CuCl	5.41	CdTe	6.46	GaAs	5.64
CuBr	5.68	AlP	5.45	GaSb	6.09
CuI	6.05	AlAs	5.63	InSb	6.45
ZnS	5.42	AlSb	6.10	SnSb	6.13
ZnSe	5.66				

APPENDIX 4
PROPERTIES OF METALS

Below are listed for various metals the valence Z, the conductivity σ, and the Hall coefficient R_H. The values are given for room temperature (300 K).

Metal	Z	σ (/$\Omega \cdot$m)	R_H (m^3/C)
Li	1	1.17×10^7	-1.7×10^{-10}
Na	1	2.38×10^7	2.50×10^{-10}
K	1	1.63×10^7	-4.2×10^{-10}
Rb	1	0.80×10^7	
Cs	1	0.50×10^7	-7.8×10^{-10}
Be	2	2.5×10^7	$+2.44 \times 10^{-10}$
Mg	2	4.45×10^7	-0.94×10^{-10}
Ca	2	2.56×10^7	
Sr	2	0.43×10^7	
Ba	2	0.17×10^7	
Nb	1	0.80×10^7	
Mn	2	0.05×10^7	-0.93×10^{-10}
Fe	2	1.03×10^7	$+0.25 \times 10^{-10}$
Co	2	1.6×10^7	-1.33×10^{-10}
Ni	2	1.6×10^7	-0.61×10^{-10}
Cu	1	5.98×10^7	-0.55×10^{-10}
Ag	1	6.29×10^7	-0.84×10^{-10}
Au	1	4.25×10^7	-0.72×10^{-10}
Zn	2	1.69×10^7	$+0.33 \times 10^{-10}$
Cd	2	1.46×10^7	$+0.60 \times 10^{-10}$
Al	3	3.76×10^7	-0.30×10^{-10}
Ga	3	0.57×10^7	

APPENDIX 4 PROPERTIES OF METALS

Metal	Z	σ (/$\Omega\cdot$m)	R_H (m^3/C)
In	3	1.19×10^7	-0.07×10^{-10}
Tl	3	0.55×10^7	$+0.24 \times 10^{-10}$
Sn	4	0.91×10^7	-0.04×10^{-10}
Pb	4	0.48×10^7	$+0.09 \times 10^{-10}$
Sb	5	0.26×10^7	
Bi	5	0.09×10^7	

APPENDIX 5

PROPERTIES OF SEMICONDUCTORS

Below are listed for various semiconductors the gap energy E_g, the electron mobility μ_n, and the hole mobility μ_p. All values are given for room temperature (300 K).

Semiconductor		E_g (eV)	μ_n (m²/V·s)	μ_p (m²/V·s)
	Si	1.12	0.135	0.0475
	Ge	0.67	0.39	0.19
III-V	AlSb	1.52		
	GaAs	1.43	0.85	0.040
	GaP	2.26	0.011	0.0075
	GaSb	0.69	0.40	0.14
	InAs	0.35	3.3	0.046
	InP	1.35	0.46	0.015
	InSb	0.16	7.7	0.075
II-VI	CdS	2.42	0.034	0.0018
	CdSe	1.74	0.060	
	CdTe	1.45	0.030	0.0065
	ZnS	3.6	0.012	0.0005
	ZnSe	2.7	0.053	0.0016
	ZnTe	2.3	0.053	0.090

APPENDIX 6
IMPURITY LEVELS IN SILICON AND GERMANIUM

All values are given for room temperature (300 K).

Donors	E_d in Si	E_d in Ge
P	0.044 eV	0.0120 eV
As	0.049	0.0127
Sb	0.039	0.0096
Bi	0.069	

Acceptors	E_a in Si	E_a in Ge
B	0.046 eV	0.0104 eV
Al	0.057	0.0102
Ga	0.065	0.0108
In	0.16	0.0112
Tl	0.26	0.01

APPENDIX 7

PROPERTIES OF SUPERCONDUCTORS

Below are listed for various superconductors the critical temperature T_c the critical field (B_c for type I superconductors and B_{c2} for type II superconductors), and the gap energy 2Δ. The values of the critical field and gap energy are given for temperature $T = 0$.

Type I	T_c (K)	B_c (T)	2Δ (eV)
Al	1.180	0.0105	0.34×10^{-3}
Ga	1.083	0.0058	0.33×10^{-3}
Hg	4.153	0.0411	1.65×10^{-3}
In	3.408	0.0281	1.05×10^{-3}
Pb	7.193	0.0803	2.73×10^{-3}
Sn	3.722	0.0305	1.15×10^{-3}
Ta	4.47	0.0829	1.4×10^{-3}
Ti	0.39	0.010	
W	0.015	0.000115	
Zn	0.85	0.0054	0.24×10^{-3}

Type II	T_c (K)	B_{c2} (T)
$Gd_{0.2}PbMo_6S_8$	14	61
Nb_3Al	18	32
$Nb_3(AlGe)$	21	44
Nb_3Ga	20	35
Nb_3Ge	23	38
Nb_3Sn	18	28
NbTi	10	15
V_3Ga	15	26

APPENDIX 8

UNITS

The SI units are given in parentheses. Other commonly used units are given in terms of the SI units. Symbols conform with the recommendations of the American National Standards Institute (ANSI) and the American Institute of Physics (AIP).

length: meter (m)
- nanometer — 1 nm $= 10^{-9}$ m
- micrometer — 1 μm $= 10^{-6}$ m
- millimeter — 1 mm $= 10^{-3}$ m
- centimeter — 1 cm $= 10^{-2}$ m
- kilometer — 1 km $= 1000$ m
- angstrom — 1 Å $= 10^{-10}$ m
- inch — 1 in. $= 2.54 \times 10^{-2}$ m
- foot — 1 ft $= 0.3048$ m
- mile — 1 mi $= 1609$ m

mass: kilogram (kg)
- gram — 1 g $= 10^{-3}$ kg
- atomic mass unit — 1 amu $= 1.661 \times 10^{-27}$ kg
- slug — 1 slug $= 14.59$ kg

force: newton ($N = kg \cdot m/s^2$)
- dyne — 1 dyn $= 10^{-5}$ N
- pound — 1 lb $= 4.448$ N

pressure: pascal ($Pa = kg/m \cdot s^2$)
- atmosphere — 1 atm $= 1.013 \times 10^5$ Pa
- pounds per square inch — 1 psi $= 6895$ Pa
- centimeter of mercury — 1 cm Hg $= 1333$ Pa

time: second (s)

nanosecond	1 ns	$= 10^{-9}$ s
microsecond	1 μs	$= 10^{-6}$ s
millisecond	1 ms	$= 10^{-3}$ s
minute	1 min	$= 60$ s
hour	1 h	$= 3600$ s

frequency: hertz (Hz $= s^{-1}$)

kilohertz	1 kHz	$= 10^3$ Hz
megahertz	1 MHz	$= 10^6$ Hz
gigahertz	1 GHz	$= 10^9$ Hz
radians/second	1 rad/s	$= 1/2\pi$ Hz

energy: joule (J $= $ kg $\cdot$ m^2/s^2)

erg	1 erg	$= 10^{-7}$ J
electron volt	1 eV	$= 1.602 \times 10^{-19}$ J
calorie	1 cal	$= 4.187$ J
kilowatt-hour	1 kW·h	$= 3.6 \times 10^6$ J
British thermal unit	1 Btu	$= 1055$ J

power: watt (W $= $ kg $\cdot$ m^2/s^3)

milliwatt	1 mW	$= 10^{-3}$ W
kilowatt	1 kW	$= 1000$ W
horsepower	1 hp	$= 745.7$ W

charge: coulomb (C $= $ A $\cdot$ s)

electric potential: volt (V $= $ kg $\cdot$ m^2/s$^3 \cdot$ A)

microvolt	1 μV	$= 10^{-6}$ V
millivolt	1 mV	$= 10^{-3}$ V
kilovolt	1 kV	$= 1000$ V

APPENDIX 8 UNITS

current: ampere (A)
- microamp $1\ \mu A$ $= 10^{-6}\ A$
- milliamp $1\ mA$ $= 10^{-3}\ A$

resistance: ohm $(\Omega = kg \cdot m^2/s^3 \cdot A^2)$
- kiloohm $1\ k\Omega$ $= 1000\ \Omega$
- megohm $1\ M\Omega$ $= 10^6\ \Omega$

capacitance: farad $(F = s^4 \cdot A^2/kg \cdot m^2)$
- picofarad $1\ pF$ $= 10^{-12}\ F$
- nanofarad $1\ nF$ $= 10^{-9}\ F$
- microfarad $1\ \mu F$ $= 10^{-6}\ F$

magnetic field: tesla $(T = kg/s^2 \cdot A)$
- gauss $1\ G$ $= 10^{-4}\ T$

magnetic flux: weber $(Wb = kg \cdot m^2/s^2 \cdot A)$
- maxwell $1\ Mx$ $= 10^{-8}\ Wb$

magnetic inductance: henry $(H = kg \cdot m^2/s^2 \cdot A^2)$
- microhenry $1\ \mu H$ $= 10^{-6}\ H$
- millihenry $1\ mH$ $= 10^{-3}\ H$

temperature: kelvin (K)
- degrees Celsius $0°C$ $= 273.15\ K$

angle: radian (rad)
- degree $1°$ $= \pi/180\ rad$
- revolution $1\ rev$ $= 2\pi\ rad$

APPENDIX 9
FURTHER READING

These are books which I have read and highly recommend for the interested student.

High School Level

Alan Holden, *The Nature of Solids* (Columbia U. Pr., New York, 1965). This book is written for the lay reader. No science background is needed.

Undergraduate Level

L. Solymar and D. Walsh, *Lectures on the Electrical Properties of Materials*, 3rd Ed. (Oxford U. Pr., New York, 1984). This book is specifically written for electrical engineering students. It is written in a lecture style and is very enjoyable to read.

J. Seymour, *Electronic Devices and Components* (Wiley, New York, 1981). This book gives lots of details about devices.

M. Ali Omar, *Elementary Solid State Physics* (Addison-Wesley, Reading, Mass., 1975). I have used this book for an upper-division course for physics majors.

Gerald Burns, *Solid State Physics* (Academic Press, New York, 1985). An excellent book with lots of physical insight. I have used this book for an upper-division course for physics majors.

John P. McKelvey, *Solid State and Semiconductor Physics* (Harper and Row, New York, 1966). This book has very good chapters on introductory quantum mechanics and statistical mechanics.

Charles W. Wert and Robb M. Thomson, *Physics of Solids*, 2nd Ed. (McGraw-Hill, New York, 1964).

M. N. Rudden and J. Wilson, *Elements of Solid State Physics* (Wiley, New York, 1980).

Adir Bar-Lev, *Semiconductors and Electronic Devices*, 2nd Ed. (Prentice Hall International, Englewood Cliffs, New Jersey, 1984).

A. C. Rose-Innes and E. H. Rhoderick, *Introduction to Superconductivity* (Pergamon, New York, 1978).

Graduate Level

Neil W. Ashcroft and N. David Mermin, *Solid State Physics* (Holt, Rinehart, and Winston, New York, 1976).

Charles Kittel, *Introduction to Solid State Physics*, 4th Ed. (Wiley, New York, 1971).

INDEX

Absolute value of complex number 116
Absorption 266
  in superconductors 289
  optical 258
Acceptor 201
Acceptor level 201
  table of 303
Acoustic branch 75, 82, 107
Amplitude 37
Angle, Bragg 47
Angular frequency 38
Anharmonic oscillation 85
Anti-Stokes shift 109
Atom, diameter of 11
Atomic mass 17
  table of 294
Atomic model of solids 166
Avalanche breakdown 243
Avogadro's number 35, 293
Band
  1s, 2s, etc. 170
  energy 158
  frequency 79
  number of states in 162
Band structure
  of barium 164
  of copper 167
  of gallium arsenide 211
  of silicon 213
  of sodium 160
Band theory of metals 153
Bardeen, J. 245, 281
Barium, band structure of 164

Base 245
Basis 23
Basis vector 3
bcc lattice 12, 298
BCS theory 281
Biased junction 231
  forward 231
  reverse 234
Bloch function 157, 158
Body-centered cubic lattice 12, 298
Boltzmann's constant 143, 293
Bond
  covalent 34
  ionic 32
  metallic 35
Boson 139
Bragg angle 47
Bragg's Law 47, 56
Branch
  acoustic 75, 82, 107
  optical 75, 82, 107
Brattain, W. H. 245
Bravais lattice 22, 30, 82
Breakdown
  avalanche 243
  reverse 243
  Zener 243
Brillouin scattering 109
Brillouin zone, first 68, 74, 79, 105, 157, 175
  volume of 162
Built-in voltage 218
Capacitance of junction 238

Cesium chloride structure 25, 298
Charge, electron 293
Collector 245
Collision time 92, 151
Collisions, electron 90, 151, 174, 180
Complex number 116
  absolute value of 116
Compton effect 100, 259
Conduction band 191
Conduction electron 89
  density of 89
Conductivity 92, 94, 147, 150, 171, 206
  table of 300
  temperature dependence of 181, 209
Constructive interference 39
Contact
  ohmic 254
  rectifying 258
Contact potential 218, 221
Cooper, L. N. 281
Cooper pairs 281
Copper
  band structure of 167
  Bragg reflections in 57, 58
  dispersion curve in 80
  Fermi surface of 167
  inelastic scattering of neutrons in 104
Core electron 89
Covalent bond 34, 200
Critical field 278
  table of 304
Critical temperature 273
  table of 304
Crystal 1, 23
Crystal momentum 102, 263
Crystal structure, table of 298
CsCl structure 25, 298
Current 90
  density 93
  persistent 274
Cyclotron resonance 97, 190
Davisson, C. J. 111
De Broglie wavelength 111
Degenerate semiconductor 263
Density
  atomic 12
  conduction electrons 89
  electron, temperature dependence of 206
  electrons in the CB 195
  holes in the VB 198
  mass 12, 17
  occupied states 145
  occupied states in sodium 164
  of states 136, 161, 193, 285
  of states in k-space 137
Density of states
  in sodium 161
Depletion layer 218, 249
  electric field in 224, 236
  width of 224, 236

Destructive interference 40
Diamagnet 275
Diameter of atom 11
Diamond, dispersion curve in 83
Diamond structure 28, 200, 298
Diatomic lattice, vibrations in 72
Diffraction
  electron 111
  grating 42
  multiple-slit 42
  neutron 112
    in iron 113
  x-ray 37, 44, 84
    in copper 57, 58
    in powdered sample 57
Diffusion 215
Diode 234, 241
  Schottky-barrier 258
Diode laser 270
Direct lattice 54
Direct-gap semiconductor 265
Directions in crystals 8
Dispersion 64, 73, 79
Dispersion curves
  in copper 80
  in diamond 83
  in potassium bromide 83
Donor 201
Donor level 201
  table of 303
Doping 201
Drain 249
Drift velocity 92

Effective mass 183, 190, 191
Einstein, A. 99
Electric field in depletion layer 224, 236
Electron
  charge 293
  collisions 90, 151, 174, 180
  diffraction 111
  mass 293
  states 137
Elements, table of 294
Emission
  spontaneous 266
  stimulated 266
Emitter 245
Equivalent positions 2, 21, 28
Exclusion principle 138
Expansion, thermal 84
Extrinsic 206
Face-centered cubic lattice 17, 298
Faraday's law 274, 275
fcc lattice 17, 298
Fermi energy 140, 195, 205, 221, 236, 290
Fermi surface 147, 166, 174, 283
  of copper 167
  of sodium 165
Fermi velocity 140
Fermi-Dirac distribution function 142, 194
Fermions 139

Field-effect transistor (FET) 249
First Brillouin zone 68, 74, 79, 105, 157, 175
 volume of 162
Flux, trapped 274
Force, restoring 62
Free particle, wave function of 116, 118, 130
Freeze-out region 206
Frequency 38
Gallium arsenide
 band structure of 211
Gap
 energy 158, 191, 287
  in semiconductors, table of 302
  in superconductors, table of 304
 frequency 79
Gate 249
Gauss's Law 224
Generation current 231
Germanium, table of impurity levels in 303
Germer, L. H. 111
Grating, diffraction 42
Group velocity 122, 131, 171
Hall coefficient 96
 table of 300
Hall effect 95, 189, 210
Harmonic oscillation 62, 84
Heisenberg uncertainty principle 124
Hole 185, 189, 191, 200
 heavy 212

light 212
Hooke's Law 62, 84
Imaginary numbers 115
Impurity level
 table of 303
Impurity state 201
Index of refraction 110
Indirect semiconductor 265
Inelastic scattering
 of neutrons 102
  in copper 104
 of photons 106
Infrared 258
Injection laser 270
Insulator 179
Interference
 constructive 39
 destructive 40
 of light waves 42
 of sound waves 39
Intrinsic 206
Inversion, population 268
Ionic bond 32
Iron, neutron diffraction in 113
Isotope effect 288
JFET 249
Junction
 abrupt 215
 biased 231
 capacitance of 238
 metal-semiconductor 254
 step 215
 p-n 215
k-space 54
Laser 266
 diode 270

injection 270
ruby 268
Lattice 2
  body-centered cubic 12, 298
  Bravais 82
  diatomic, vibrations in 72
  direct 54
  face-centered cubic 17, 298
  monatomic, vibrations in 63
  non-cubic 30
  parameter 8
    table of 298
  reciprocal 51
    one-dimensional 71
  simple cubic 6
  three-dimensional, vibrations in 79
  vector 3
  vibrations 61, 180
  waves
    in copper 80
    in diamond 83
    in potassium bromide 83
Light
  speed of 293
  visible 258
Light meter 260
Light scattering, inelastic 106
Light-emitting diode (LED) 262

Longitudinal wave 65, 82, 109
Magnetic field 95, 97, 189, 190, 210, 274, 276
  critical 278
Maiman, T. 268
Mass
  density 17
  electron 293
  neutron 293
  proton 293
  table of atomic 294
Mass, atomic 17
Matter wave 115
Meissner, W. 275
Meissner effect 275
Metal-semiconductor junction 254
Metallic bond 35
Metals
  band theory of 153
  classical model of 89
  free-electron model of 133
  table of properties of 300
Metastable state 268
Miller indices 11
Mixed state 280
Mobility
  electron 208
  hole 208
  in semiconductors, table of 302
  temperature dependence of 209
Monatomic lattice, vibrations in 63

MOSFET 251
Multiple-slit diffraction 42
NaCl structure 21, 299
Neutron diffraction 112
   in iron 113
Neutron mass 293
Neutrons
   inelastic scattering of 102
   in copper 104
Ochsenfeld, R. 275
Ohmic contact 254
Onnes, K. 273, 278
Optical branch 75, 82, 107
Orbital 166
Oscillation
   anharmonic 85
   harmonic 62, 84
p-n junction 215
   biased 231
Partial derivative 117
Particle in a box 125, 133
Pauli's exclusion principle 138
Penetration depth 276
Period 38
Periodic boundary condition 135
Permeability constant 281, 293
Permittivity constant 32, 224, 293
Phase velocity 122
Phonon 101, 265, 269, 282
   absorption 102
   emission 103
Photoconductive cell 260
Photodiode 261

Photon 99, 258
   inelastic scattering of 106
Photosensitive devices 260
Phototransistor 262
Physical constants, table of 293
Pinching 251
Planck's constant 99, 293
Planes in crystals 8
Polarization of lattice wave 79
Population inversion 268
Potassium bromide, dispersion curve in 83
Potential, contact 218, 221
Primitive unit cell 12
Probability function 118
Proton mass 293
Quantized
   electromagnetic wave 99
   lattice wave 102
Quantum 99
Quantum mechanics 115
Raman scattering 107
Real numbers 115
Real space 54
Reciprocal lattice 51
   one-dimensional 71
Reciprocal space 54
Recombination 200
Recombination current 228, 241
Rectifier 234, 241
Rectifying contact 258
Refraction, index of 110
Resistance 94
Resistivity 94

    of sodium, temperature
        dependence of 182
Restoring force 62
Reverse breakdown 243
Ruby laser 268
sc lattice 6
Schottky-barrier diode 258
Schrieffer, J. R. 281
Schroedinger's equation
    117, 130
Semiconductor 180, 191
    degenerate 263
    devices 241
    direct-gap 265
    II-VI 200
    III-V 200
    indirect-gap 265
    table of properties of 302
    n-type 203
    p-type 203
Silicon
    band structure of 213
    table of impurity levels
        in 303
Simple cubic lattice 6
Sodium
    band structure of 160
    density of occupied
        states in 164
    density of states in 161
    Fermi surface of 165
    temperature dependence
        of resistivity of 182
Sodium chloride structure
    21, 299
Solar cell 262

Sound, velocity of, in crystals 68, 75
Source 249
Space-charge region 220
Speed of light 293
Spin states 137
Spontaneous emission 266
Stimulated emission 266
Stokes shift 109
Superconducting magnet
    278, 280
Superconductivity 273
Superconductor
    table of properties of 304
    type II 279
Symmetry, translational 2
Temperature dependence
    of conductivity 209
    of electron density 206
    of mobility 209
    of resistivity of sodium
        182
Thermal
    energy 61, 86
    expansion 84
Thomson, G. P. 111
Three-dimensional lattice,
    waves in 79
Transistor 245
    p-n-p 245
    bipolar junction 245
    field-effect 249
    metal-oxide-
        semiconductor 251
    n-p-n 245
Translational symmetry 2
Transverse wave 66, 82, 109

Tunneling 128, 243, 290
Ultraviolet 258
Uncertainty principle 124
Uncompensated electrons 150, 175
Unit cell 6
  primitive 12
Units, table of 305
Valence 90
Valence, table of 300
Valence band 191
Valence electron 89
Velocity
  group 122, 131, 171
  of light 293
  of sound in crystals 68, 75
  of wave 38
  phase 122
Vibrations, lattice 61, 180
  diatomic 72
  monatomic 63
  three-dimensional 79
Visible light 258
Voltage, built-in 218
Volume of first Brillouin zone 162
Vortex state 280
Wave 37
  function 115
  in diatomic lattice 72
  in monatomic lattice 63
  in three-dimensional lattice 79
  longitudinal 109
  transverse 109
  vector 52, 79
  velocity 38
Wave function of a free particle 116, 118, 130
Wave number 38
Wavelength 37
Wigner-Seitz cell 12
Work function 254
X ray 45
  diffraction 37, 44, 84
    in copper 58
    in powdered samples 57
Zener breakdown 243
Zener diode 245
Zincblende (ZnS) structure 26, 200, 299